# PLACES OF LAST RESORT

# Places of Last Resort

*The Expansion of*
*the Farm Frontier*
*into the Boreal Forest in Canada,*
*c. 1910–1940*

J. DAVID WOOD



McGill-Queen's University Press
Montreal & Kingston · London · Ithaca

ISBN-13: 978-0-7735-3039-3 ISBN-10: 0-7735-3039-8

Legal deposit second quarter 2006
Bibliothèque nationale du Québec

Printed in Canada on acid-free paper

This book has been published with the help of a grant from the Canadian Federation for the Humanities and Social Sciences, through the Aid to Scholarly Publications Programme, using funds provided by the Social Sciences and Humanities Research Council of Canada.

McGill-Queen's University Press acknowledges the support of the Canada Council for the Arts for our publishing program. We also acknowledge the financial support of the Government of Canada through the Book Publishing Industry Development Program (BPIDP) for our publishing activities.

---

**National Library of Canada Cataloguing in Publication**

Wood, J. David (John David)
Places of last resort : the expansion of the farm frontier into boreal forest in Canada, c. 1910-1940 / J. David Wood.

Includes bibliographical references and index.
ISBN-13: 978-0-7735-3039-3 ISBN-10: 0-7735-3039-8

1. Agriculturally marginal land – Canada – History – 20th century.
2. Agriculturally marginal lands – Government policy – Canada.
3. Frontier and pioneer life – Canada. 4. Clearing of land – Canada – History – 20th century. I. Title.
S451.5.A1W66 2006 333.76'0971'09041 C2006-900193-6

---

Typeset in Palatino 10.5/13
by Infoscan Collette, Quebec City

*For Mary, Lisa, Stepan, and Evan*

# Contents

# List of Figures, Tables, and Illustrations

FIGURES

TABLES



ILLUSTRATIONS

# Preface

Being a farmer was still the classic honourable livelihood in Canada in 1910, but the search for farmland was approaching "last call." Without the embellishments that F.J. Turner added to the idea of the frontier, the expectation of the ongoing expansion of farmland had a powerful grip on the popular imagination, and the hint of an end to the agricultural frontier was an unwanted and somewhat sinister rumour. As Kenneth Boulding recognized in framing his influential Spaceship Earth concept, "over a very large part of the time that man has been on earth, there has been something like a frontier. That is, there was always someplace else to go when things got too difficult ... The image of the frontier is probably one of the oldest images of mankind, and it is not surprising that we find it hard to get rid of."[1] North America and the acquisitive Europeans who came to claim it had enjoyed three centuries of almost unstinted availability of cheap land and relatively unrestricted access to economic opportunities. The ending of the agricultural frontier, though not recognized at the time, was a striking early expression of the limits inherent in the closed system Boulding called Spaceship Earth. New land for farming was a fundamental example of an exhaustible resource necessary for the survival of the human species, and its sudden scarcity gripped more than the farming community. The running out of new land was only the first example; it has been followed by other examples of near-exhaustion or debasement of important, life-sustaining resources that are vital to urban as well as rural lives.

Many settlers were gripped by an urgency to find a farm in the face of the threatened scarcity. As with any frontier, the composition of the settler population was very variable, with a portion being knowledgeable and competent and a portion being unsuited for the challenges of life at the edge. Whatever their individual qualities, the settlers were unwittingly engaged in demonstrating for the society as a whole the real limits of the land available for an agricultural frontier. Across the three thousand kilometres of the boreal forest margin addressed by this book, there were a few areas where the demonstration was positive, but many more where the result was not enough for a decent living. The process typically involved extraordinary demands on physical strength and tolerance of financial uncertainty, probably more than characterized the favoured nineteenth-century stages in the expansion of agriculture across the continent. The north was a genuine successor to the drought-prone grassland frontier in optimistic expectations, but quite different in characteristic crops and farming techniques. The boreal expansion opened land hundreds of kilometres north of the subhumid areas where the grassland frontier was coming to a halt. The move north, unlike its predecessor, was *national* rather than regional in scope, through its extension into Ontario, Québec, and British Columbia. The boreal forest was generally a difficult frontier, and it was being entered at a time of heightened concern over the rural way of life in Canada. The importance given to finding new farmland in the three decades leading up to the 1930s had little effect on the relentless shift of population from rural to urban areas. The exodus even from successful rural areas was under way and gathering momentum during this period, and many settlers in the boreal margin were joining the flow less than one generation after settlement.

Leaving has continued to be a major feature of the last attempts to maintain an agricultural frontier. It is not uncommon to meet someone in southern Canada who spent a childhood in the north or has family members who tried to make a farm there. Comparable stories could be gathered in other parts of the world where zealous searches for farmland were going on after the opening of the twentieth century. In the English-speaking world, most notably in British dominions, farming expanded into inland New South Wales, the mallee in Victoria, the Eyre Peninsula in South Australia, and the southwest section of Western Australia; and in New Zealand into forested hill country, to name the most substantial examples outside North America. All these late-settled farming areas, representing a significant portion of Isaiah Bowman's "Pioneer Belts," have proven to be major generators of mobile population: feeling marginalized seems to provide an impetus to move and to try to improve one's status. This is an age-old response, as illustrated by fringe areas of eastern Europe that over the centuries have given rise to outbursts of enterprise and innovation.[2] At the opening of World War II, the exodus from Canada's boreal margin burst into full flood, marking the end of the long period of almost continuous expansion of agricultural land in northern North America.

This book is the story of the boreal forest farm settlement in Canada. In addition to recognizing the passion for land of the many individual farm seekers, this study illustrates the influences of various agencies and their policies in encouraging settlement beyond previous limits. The primary precursor for this study was the Canadian Pioneer Problems Committee, which carried out its investigations on the eve of and during the Great Depression, under the advice and encouragement of Isaiah Bowman, and published a range of significant analytical volumes. The present book picks up the debate over the wisdom and method of expanding agricultural settlement, and assesses the distribution of responsibility for the tribulations experienced by settlers in marginal areas. It is shown that while farm-related scientific information, available at least to officials, increased rapidly in the early decades of the twentieth century, officials made only piecemeal efforts to apply it to the expansion under their administration. The evidence comes from many agents and individuals involved in the settling, sometimes (especially in chapter 5) in their own words, in voices that have seldom been heard. Some sources that have arisen since the 1930s, such as long climatic records, soil tests, and the Canada Land Inventory maps, are used to fill out the reconstruction of the physical environment of marginal areas. The illustrations provide a visual touchstone for the conditions in which marginal settlement was attempted in the inter-war period.

In some respects this project began with a generous grant from the Social Science and Humanities Research Council of Canada (SSHRC), 1994–97. My collaborator on that grant was Prof. Brian Osborne of Queen's University, and his engagement with the significance of marginal areas for adjacent settled society has provided one of the stimuli for the present inquiry. I am also thankful to SSHRC for a research grant in 1998, as well as for travel grants in 1998, 1999, and 2003, to allow me to deliver conference papers on aspects of the book. Many of my ongoing research expenses in the past few years have been defrayed by the Research Committee and by the office of the Dean of Atkinson College (now the Atkinson Faculty of Liberal and Professional Studies), York University. I am indebted to many individuals and research institutions. For both textual and photographic resources I am grateful to the archives of the provinces of Ontario, Saskatchewan (Saskatoon and Regina), and Alberta, as well as the National Archives of Canada in Ottawa. I also benefitted by researching in the Glenbow Museum and Archives, Calgary; the University of Saskatchewan Archives, Saskatoon; and the Archives of York University. Dr Douglas McManis made it possible for me to investigate the beginnings of the Pioneer Belts initiative at the archives of the American Geographical Society in New York. My work on the Great Clay Belt of

Ontario was generously assisted by Curator Julie Latimer of the Ron Morel Memorial Museum, Kapuskasing, and by Town Clerk Demers in the Cochrane municipal office (specifically regarding Cochrane and Glackmeyer Township). The main source for secondary material for this study has been the York University library where, especially in the Interlibrary Loan and the Map departments, research support has been provided knowledgeably and expeditiously. An ongoing significant repository for me has been the Toronto Reference Library; another has been the library of the Archives of Ontario and, for its Canadiana material, the Barrie Public Library.

I have had very competent research assistants in the persons of Charlotte McCallum, Robert Davidson, and Richard Anderson. The cartography was ably executed by Carolyn King, York University, and (figures 4.1 and 4.2) Janet Allin. John Dawson expertly carried out varied photographic operations. I had the advantage of two generous colleagues, Dawn Bowen of Mary Washington College and Carl Tracie of Trinity Western University, who agreed to read a version of the book. Both of them are recognized researchers on boreal settlement and provided excellent critiques that I trust I have done justice to in the final version. Among other scholarly benefactors, John Warkentin read an early part of the book and gave advice on things western, Ray Ellenwood edited my translation of Joseph Laliberté's account, Bill Westfall pointed out a source on the elusive Burnsites (the denomination of the so-called Bull Outfit of the Peace River country), and Stan Trimble reminded me of the work of Kirk Stone on Alaska and Scandinavia. Other friends of the project include Micheline Courchesne (*Terre de Chez Nous* newspaper), and John Becker (local and ethnic history and genealogy). The staff at McGill-Queen's University Press were typically helpful, specifically Joan Harcourt in Kingston, Joan McGilvray in Montreal, and thoughtful copy editor, Lesley Andrassy. My family has observed from the wings as the book materialized, sometimes providing quick research help or opinion on phraseology, sometimes just providing diversion; and my wife transcribed from tapes the Kerpan and Kinderwater interviews, periodically provided helpful critiques, and was always a source of intelligent perspective. The dedication is to my family, a small recognition of their contributions and of "being there."

# PLACES OF LAST RESORT

# 1

# The Impulse to Expand the Farm Frontier

If you will get a homestead map from the Natural Resources Intelligence Branch, of the Department of the Interior, you will find an occasional white strip in among the settled parts. *These lands are however of inferior quality or they would have been homesteaded long ago.*[1]

Being a pioneer was a gamble almost anywhere on the late agricultural frontier in the New World. All but the arrogant would be worried by questions: Would frost or drought destroy the crops? Would injury or disease strike down the primary workers or the essential draught animals? Would the farm prove to be unproductive? Would all members of the farm family stay committed as time passed? In many cases the answers to such questions meant failure. Unproductive land was being identified and abandoned in many places by the end of the nineteenth century. In Ontario, the so-called "waste lands," cleared and farmed sixty or more years earlier, were identified and even roughly mapped[2] and farm folk were demonstrating their assessment of expanses of poorly drained land in the Ottawa valley and the Huron Uplands by leaving in large numbers for the West. Something similar had happened in Québec, leading to the lamented exodus of population to New England and elsewhere, once settlement had expanded out of the fertile land of the St Lawrence valley.[3] Even the West was not a land of unqualified success. The cancellation of land grants, commonly meaning abandonment by settlers, had occurred not only with individual settlers but also under organized colonization schemes, as in the so-called Sowden Settlement in southwest Manitoba, where a full 15 per cent of the 720 quarter sections were cancelled and given up at least once by colonists in the 1880s.[4] Friesen recalls Chester Martin's rough calculation that four in ten prairie homesteads were not "proved up" by the applicant and goes on to demonstrate that there was a social component, beyond the quality of the land, in the recipe for success in pioneer settling.[5] Cancellation of homestead entries continued at variable rates in all kinds of areas as settlement flowed westward. Pioneering was not a ready-made job, but required making something out of the difficult raw materials of new land and climate.

If the nineteenth-century agricultural expansion could be seen as a gamble for a settler, then being a settler seeking farmland after the first decade of the twentieth century could be thought of as a much bigger, almost ill-fated gamble. As Auld pointed out to the Bruce Preparedness League in 1918 (chapter header, above), if any good land were indicated by the map, it was almost certainly claimed already. Carl Dawson and Eva Younge, in their book on the Prairie Provinces, identified what were "chronic fringes" of settlement by the 1920s, including the rough land of Riding Mountain, along the western provincial boundary of Manitoba, and the arid lands near the southeast boundary of Alberta, and reaching into the Great Sand Hills of Saskatchewan (see place names in figures 1.1 and 1.2).[6] In other areas of overseas European colonization it was difficult to find remnants of good land. Griffith Taylor claimed that in Australia virtually all the land suitable for crop farming had been occupied by the end of the nineteenth century; he argued that "almost the whole of 'economic Australia' was fairly well known sixty years ago," and there had been no successful expansion of farming in recent decades into what J.W. Gregory had called "The Dead Heart of Australia."[7] In America the reading public knew that new farmland free of serious

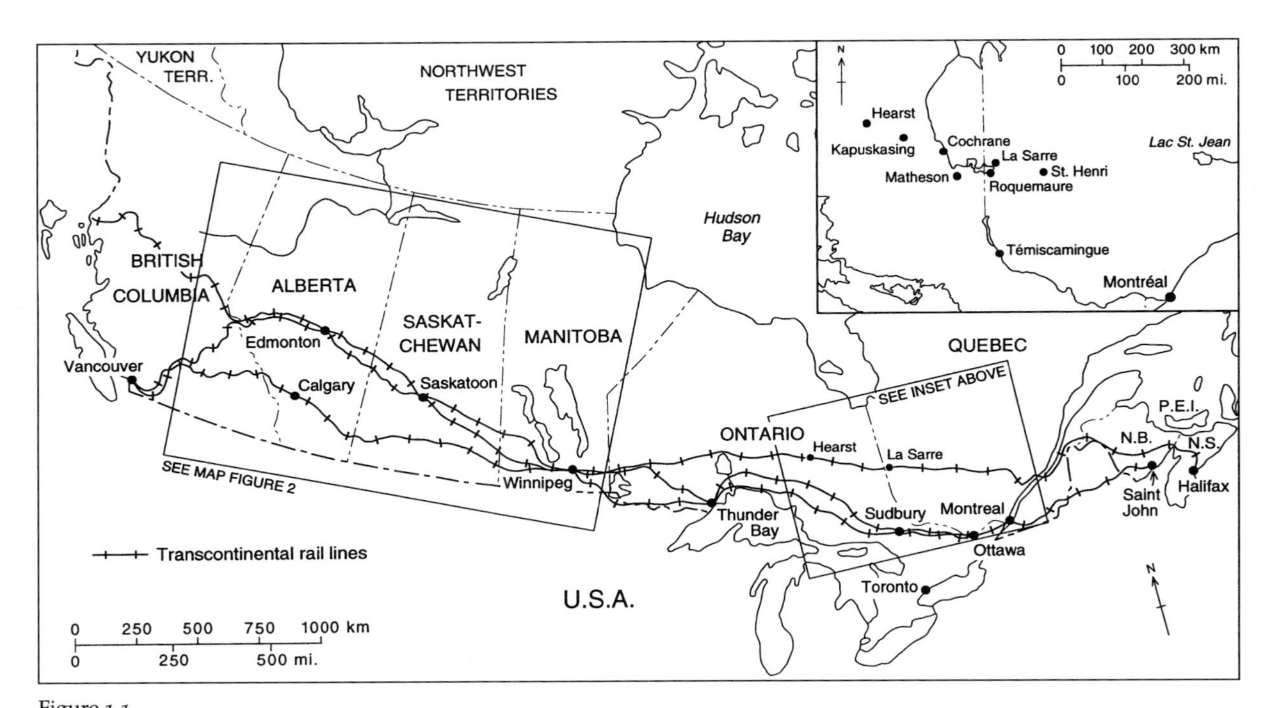


Figure 1.1
Central Canada, including major railways and provincial boundaries.

hazards was getting very scarce: the *Century Magazine*, reporting on "the last frontier" in 1909, said "we are within sight of the end of free land ... the last West has ... been reached ... among the very latest of the newly come homesteaders, it was a continual shock to find out how little really excellent land remained for free homesteading ... we have come to believe free homesteads, like the poor, we should always have with us ... Canada's free lands extend to the Pole all right; only they are not farm-lands ... when you go seventy miles north of the Saskatchewan [River], arable land exists only in small patches. The rest of the North Country is sand, muskeg, rock."[8]

When the Social Science Research Council of the United States was looking for active "pioneer belts" to study in the 1920s, it turned to other countries. Canada could provide evidence of agriculture expanding into fringe areas. One expansion was into southern parts of the Prairie Provinces previously deemed too dry for cultivated crops. This testing of the limits of reliable moisture became involved in the destructive conditions of the infamous Dust Bowl that reached a nadir in the 1930s. This book does not dwell on the well-publicized dry margin, but some aspects of its troubled career are explored to illustrate an important part of the context within which the desperate search for land was taking place. The other common agricultural expansion in Canada turned away from dryland opportunities toward northern areas where the major hazard usually was unpredictable, crop-damaging frost. This expansion was into the trees and scrub of the boreal forest, typically along a discontinuous front. This book attempts to shed light primarily on this boreal margin, from northeastern Québec to the Peace River country, and on the efforts across Canada to make it into farmland. The term "margin" is applied to these two areas of late agricultural expansion, meaning the fringe of farming territory where conditions for successful agriculture are in question. Thus it is synonymous with the normal meaning of agricultural frontier, in the sense of the leading edge of farming settlement pushing into territory

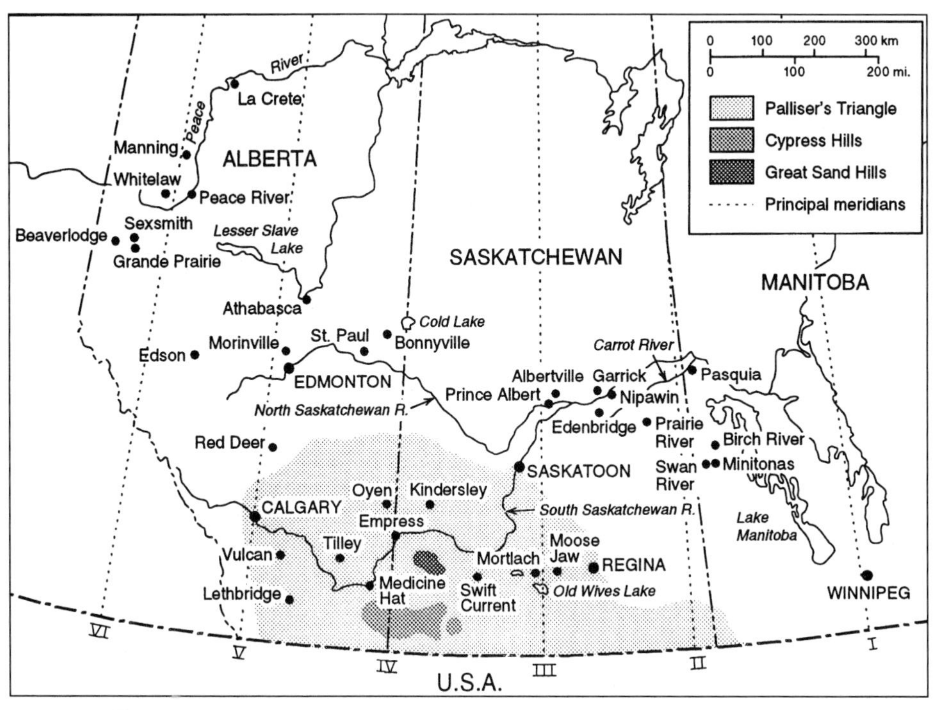


Figure 1.2
Place names and certain physical features in the Prairie Provinces (inset from fig. 1.1).

that had not been tried for tillage crops, and is sometimes interchanged with it (see figure 1.2; and for the boreal "leading edge," see figure 2.1). Distinctions among the marginal areas are discussed further in chapters 2 and 3.

Settlers invaded northern wooded land from coast to coast in Canada during the first half of the twentieth century. In New Brunswick intending farmers were encouraged to take up land in the northern counties where forestry had been the principal activity, and in the 1930s the main thrust was in assisted colonization. Similarly, Nova Scotia saw some expansion into fringe areas, usually made available by the clearing of commercial forest.[9] In Québec the agricultural college at Ste Anne de la Pocatière, on the St Lawrence River, was active in providing training for young farmers who went out to various peripheral parts of the province that seemed eligible for farming. But the major invasions of the boreal margin were from Saguenay-Lac St Jean, where settlement was spreading beyond the established farming of the late nineteenth century, across the Abitibi and Ontario "clay belts," on the north side of the parkland in Manitoba, Saskatchewan, and Alberta to the Peace River country. This huge stretch, crossing the central two-thirds of Canada, from Québec to the northeast edge of British Columbia, is the focus for the main body of the book.

## PUSHING TOWARD THE LIMITS

Before the end of the nineteenth century perceptive scholars were predicting dire consequences of the approaching end of the New World frontiers. The argument, echoing earlier European theorists, was that the availability of cheap farmland and a free-wheeling social context on the frontier had given rise not only to an economic bonanza but also to an unparalleled egalitarianism, the fundamental characteristic of American democracy.[10] In this view the ending of the agricultural frontier that had opened America to unrestricted occupation by "honest yeomen" worried thoughtful observers: it seemed that the unwelcome alternative was a gradual reversion to a class-ridden Old World society.[11]

The belief that the frontier, having stimulated the nineteenth-century boom, had to be kept alive to maintain social well-being had a powerful grip on popular thinking in the first half of the twentieth century. Influential arguments took the form of The Doctrine of Success or The Idea of Progress, a theme of which was "every day, in every way, I am [or, one could say, life is] getting better and better" (Dr Coué); or as the Rev. Norman Vincent Peale was to phrase it later in an extravagant outburst, "Through the help of God, through courage, character, manliness, and the power of positive thinking, you can make your life whatever you want it to be."[12] Positive thinking was prescribed as a replacement for the panoply of opportunities that had characterized the nineteenth century. The unique influence of positive thinking to heal all society's ills could only have been propounded where an elasticity of opportunity, as expressed in the century-old admonition "Go West Young Man," was universally accepted as the normal state of affairs. The majority of people born in the New World toward the end of the nineteenth century were nurtured by this understanding. In reviewing the lessons of the previous American century, Charles Beard opined that the "idea of progress," in contrast to the pessimism of the Old World, could emancipate mankind "from plagues, famines, and social disasters, and subjugate the materials and forces of the earth to the purposes of the good life – here and now."[13] It is not surprising then that the huge slaughter of World War I had to be explained away as being the war to end all wars, and that a renewal of the agricultural frontiers could be embraced as a way to find again the old optimism; and perhaps opening new frontiers for restless land-seekers could siphon off some of the hostility that had built up in the Old World.

The flurry of activity and debate that went into the search for and the occupation of new (or "late") frontiers from the 1910s to 1930s raises the question: What vaunted ambition or desperate need would have driven people to venture into the hitherto avoided or abandoned marginal lands? A dynamic seemed to be arising from two nearly subliminal enthusiasms feeding off one another: the one was the deep desire of people for a farm of their own, whether from the expectations in a traditional farm family or as a salvage attempt following failure elsewhere; the other was a passionate though amorphous belief among politicians and bureaucrats that the expansion of farm settlement must continue. Some accepted Turner's claim that the spreading frontier was the essential element in the American way of life, and a major influence in the world view of Canadians; others feared losing what had become the entrenched – and successful – way of doing things in the New World. But it was not far into the twentieth century before people began to move into land that had not been intended for plough farming, that had not previously been surveyed into the characteristic quarter sections. Decisions by ambitious politicians made it possible for often unprepared settlers to move into areas that had been set aside as forest reserves, or had been run as ranches for decades, or were remote Crown (publicly owned) land. The cases of failure and abandonment that followed, beginning as early as the end of the 1910s, had sociopolitical repercussions: they eroded little by little the brash optimism that had been nurtured by the rapid expansion of New World agriculture and industry in the nineteenth century. And the widely voiced suspicion that good land for farm crops was no longer to be found, despite eager searches and the opening of new areas, raised the disturbing possibility that the frontier and its fabled influence were virtually at an end.

Renewal of the frontier was an aim of the Social Science Research Council in the United States, even though its members knew that virtually no remnants of the frontier would be found internally. One of the Council's earliest projects, in 1927, was "A Scientific Study of Settlement in Pioneer Belts." This project was largely inspired and promoted by the geographer Isaiah Bowman, and the roster of practical studies to implement the project was designed at the outset principally by the agricultural economist O.E. Baker. Bowman had been developing this project for some time, and its main themes, probably arising from his experiences as part of the United States delegation at the Versailles peace negotiations, had been laid out in an article published in 1926.[14] It seems likely that Bowman thought land hunger was one of the causes of the war, and one that had not been ameliorated. O.E. Baker, through his position in the United States Department of Agriculture, had been involved in the preparation of an atlas of the northern plains states, focussing on land classification for agriculture.[15] Well-informed scholars, such as Bowman and his colleagues in the Social Science Research Council, recognized that, as settlement began pushing beyond the established peripheries of farming, mistakes became much more likely. Of course, mistakes had not been rare on earlier frontiers, where failures or strategic withdrawals accounted for a significant proportion of those who tried. But, as even farther limits were approached, the twentieth-century failure and extreme hardship loomed much larger. The intention of the Pioneer Belts committee was to sustain the attraction of the frontier experience, and its supposed sociopolitical benefits, by providing a template for successful marginal settlement that could be applied by governments in various places: a scientific approach based on a checklist of required preparations would ensure success.

Bowman's various expositions of a science of settlement, between 1926 and 1937, in combination provide the basic components. Fundamental to success is the involvement of government 1) to help plan and finance the settlement; 2) to commission the preparation of accurate and up-to-date information on climate, soils, and related natural characteristics; and 3) to assist experimentation with crops and techniques to determine the best farm regimen for the conditions. It is also necessary 4) for applicant settlers to be prepared to follow a different farming livelihood than the one familiar to them. In the "modern" era, 5) the provision of at least basic modern amenities is a requirement for a stable settlement; the deprivations endured by the "old-time" frontiersmen would not suffice. Bowman introduces some social science as well, recommending 6) that the settlers destined for the new land should be deemed culturally and mentally suited to the challenges of the target area.[16] Notwithstanding the availability of much more information about the natural environment and about ways of expediting settlement, the application of "scientific settlement" by government agencies, while somewhat more common than in the previous century, was usually piecemeal and not rigorous (as illustrated in later chapters).

Governments had been encouraging expansion of settlement in New World frontiers through much of the nineteenth century. In the United States the vigour of the expansion was such that easterners were filing into valleys near the Pacific coast as early as 1850, when the Willamette and neighbouring valleys contained about 12,000 souls.[17] The opening of the twentieth century gave a fresh impetus to finding new frontiers. In Canada, for instance, farming was moving into the clay belts of northern Québec and Ontario and into the Peace River country by the end of the first decade. About the same time in Australia, settlers were grappling with the mallee in western Victoria, making incursions as far as the western boundary, and in South Australia were pushing ahead of the railway into the dry margins of the Eyre Peninsula; slightly later, wheat farming started to expand toward the coast southeast of Perth in Western Australia. In the United States, O.E. Baker reiterated the opinion heard twenty years earlier, that no new agricultural territories were to be found (at least in the contiguous forty-eight states): a north-south strip of the Great Plains had been the only significant line of frontier expansion in the first three decades of the twentieth century, and it was "the last agricultural frontier in the United States."[18] There was hope for the expansion of agriculture in the wooded lands of Alaska, but only in a few areas of favourable climate could potatoes, spring wheat, barley, or oats be produced.[19] Settlement was expanding into the southwest of the Matto Grosso in Brazil, and the adjacent Gran Chaco of Bolivia, Paraguay, and northern Argentina. In South Africa what appeared to be its last pioneer belt was being established in the northern Transvaal. The building of the Trans-Siberian Railway in Russia was serving to siphon settlers in large numbers toward the far reaches of eastern Siberia.[20] The various schemes that enticed settlers beyond previous limits in the 1920s and 1930s were in most cases expressions of the zealous encouragement of governments, in the belief that growth of the rural population was necessary for economic well-being, let alone political stability. The expansion of rural settlement, providing for the continuation of familiar traditions, was taken to be almost a divine imperative. Warnings about the limitations of the margins fell on the deaf ears of the champions of growth who were imbued with the expansionist beliefs and rhetoric that had influenced their formative years. Griffith Taylor, in sparring with the champions of growth in Australia over the potential for agricultural expansion, described them as "'patriots' who believe that nothing but praise is permissible about their native land" and its potential for continuing to offer the pioneering dream.[21]

## CHALLENGING THE DOCTRINE OF SUCCESS

During the nineteenth century few questioned the potential for success on the agricultural frontier. Although there had been setbacks, such as the petering out of good land for homesteading in Ontario in the 1850s, or the drought in South Australia in the 1860s that led to the famous "Goyder's line" barring settlers from arid land, they had been surmounted and replaced by a range of new opportunities. Pleas for preservation of part of the natural environment that was being obliterated for the purposes of agriculture were viewed as rantings, and their authors, the most famous of whom in America was John Muir, were marginalized. Before the twentieth century had advanced more than a few years, however, questions began to be raised both about the kind of land that settlers were being encouraged to occupy and about the suitability of the laisser-faire process under which frontier settlement had traditionally been pursued.

In Ontario efforts were renewed to put settlers into the northern half of the province, especially adjacent to the railway lines that were being extended (see figure 1.1). The railways passed through what had come to be called the clay belts – areas overlying the Precambrian Shield where post-glacial lakes had left lenses of mixed friable and coarse soil material on which a heavy boreal forest had grown. Once the trees had been removed a checkered terrain of workable land remained, but for most farm crops there were too few "growing degree days"[22] and too short a frost-free season. Yet settlers were encouraged to move into these areas, in both Québec and Ontario.

Two economic activities were proposed for the clay belts: farming and harvesting the vast forest for pulpwood, sawlogs, and fuelwood. This was to be a renewed forest frontier such as had earlier passed through the forests of the Eastern Seaboard, New England, the St Lawrence valley, and southern Ontario. The result of the earlier frontier had been a productive and prosperous agricultural economy, so there was hope for the clay belts. The federal Commission of Conservation undertook to assess the potential of areas within easy reach of the railway that crossed northern Ontario and for the purpose engaged the dean of the new School of Forestry at the University of Toronto, B.E. Fernow. His survey was restricted to a few hundred metres north and south of the railway line, embellished by information from a survey six years earlier and from occupants of the region, and was completed in about a week in October 1912. Fernow's mandate was to assess the quality of the timber and of the land for agriculture, and in so doing he challenged the optimism of the civil servants who had named the area "New Ontario." Although he admitted that his judgment "runs counter to many other opinions as to the value of this part of the clay belt," he noted, with reference to both forestry and agriculture, that the most impressive feature of the area was swamp.[23] Even the so-called clay belt soils varied in physical and chemical characteristics and, in many areas near the rail line, had been burned over. He reiterated the judgment of the 1906 analysis that there was no certainty about the area supporting an agricultural population. Persuaded by scientific caution rather than boosterism, he concluded "that probably 50 per cent of the area involved does not contain any wood values, and that probably the same percentage of it is, *under present conditions*, undesirable to open for settlement ... more systematic and careful direction of settlement is highly desirable."[24]

At this time, while the available land of the Prairie Provinces was rapidly filling up, business growth began to level off, and the year 1913 was marked by a precipitous drop of the real-estate market in Edmonton, repeated in other urban areas. Rural areas continued to attract settlers after the hiatus of the World War, thanks to federal government publicity. In contrast to the eastern provinces, the Crown lands in the Prairie Provinces, namely the lands eligible for and involved in homesteading, were controlled until 1930 by the federal government, to the perennial dissatisfaction of the provincial governments. Thus the Saskatchewan bureaucrat, Auld, in offering advice in 1918 to the potential settlers in Walkerton, Ontario, spoke openly about the lack of good homesteads (see the quotation opening this chapter).[25] Informed persons understood that land suitable for the kind of farming that had been perfected over the centuries in northwestern Europe and eastern North America was no longer to be found.

The advice offered to the potential settlers in Ontario would have rung true for Wilfrid Eggleston's father who, as an immigrant from England, was keen to find a homestead. The location he chose in 1910, at the urging of a friend who would be his neighbour, was in one of the driest parts of the western interior, near the southeast corner of Alberta. But the area had been opened for homesteading on 160-acre parcels, and a movement of people, wagons, and animals was under way. Although Eggleston did not bring his young family out to the property for two years, his decision was not to be changed, and neither was the outcome thirteen years later when the family abandoned the place along with 90 per cent of the homesteaders. Musing on the experience sixty years later, Eggleston wonders "whether my father faced up to the fact that even in this marginal area our own homestead had to be regarded as one passed up in the early land rush ... Otherwise how did it happen that it was still open for his entry five months after the first claims were filed? ... even the ranchers had stayed away from those arid townships until all the more attractive ranch sites in southern Alberta had been occupied."[26]

On the far northwestern frontier, near the Alberta border with British Columbia, W.D. Albright was coming to terms with his own homestead, with the newly opened region, and with the whole process through which pioneers had been enticed to more and more remote fringes. Albright was a student of frontier settlement and, having been a newspaperman in Ontario, he was equipped to write about it. In addition, he soon turned his property near Beaverlodge, Alberta, into an experimental farm, and set about analysing the region as farming territory. In a cryptic note at the Beaverlodge farm, he recorded the experiences of a few Peace River farmers in growing wheat in the early, frost-plagued years of settlement. With reference to one hard-hit farm near Whitelaw in the North Peace River area, he wrote: "I have marked 1918 1922 & 1923 as poor [on] acc't [of] frost. 1918 almost total loss. 1922 reported only about 5 days threshing. 1923 frost in some districts made total loss but was not a total loss [in] all [of] one district."[27] A more tragic example from the neighbourhood of Beaverlodge was of someone who came late in the search for land, influenced by what Albright called "homesteaditis":

Twenty-three years ago a lumberjack homesteaded on a high hill-top six miles north-east of the Beaverlodge Experimental Station. It was reached by a break-neck trail and once the family got in the wife seldom got out. Little crop could be raised but potatoes and vegetables. Three years the family lived on that lofty look-out ... During those three years the family of five or six children never went to school. One girl died ... Their ill-fated venture proved a tragedy. In their case it was not so much distance as inaccessibility and unsuitable land.[28]

Success was proving elusive in relatively new settlement areas in other parts of the world. In Australia, legislation was encouraging closer settlement through a process of breaking large pastoral estates into pieces to be sold to crop farmers. By the end of World War I, when soldiers were returning to find farms, closer settlement was encroaching on the arid margins.[29] Certainly by 1930 Griffith Taylor's claim of the limits to traditional crop farming was again being proven right: the Dead Heart would periodically reassert its destructive influence. New Zealand provides many other examples, especially in the discharged soldiers' settlements, of the attempts to spread agricultural occupance in New World countries after World War I.[30]

## A FIRST GLIMPSE OF THE BOTTOM OF THE BARREL

The twentieth century proved to be a time of learning some bitter lessons: from its opening in a cloud of optimism, through a gradual realization of an overcrowded planet, highlighted by two huge conflicts and many lesser ones, it finished with struggles over most fundamental resources and recognition that basic conditions for human life on earth – whether air, water, or food – are threatened with poisoning. Getting to this understanding has been almost an osmotic process for our society: stumbling from one crisis to another – from shocking measurements of the extent of the chemical contamination of the environment, to the tallying of the rate of extinction of animal and plant species, to threats to the supply of essential fuel – society has reluctantly begun to accept that we have to do a better job of caring for the world we rely on. Early halting steps toward acceptance were taken in the 1930s by a rural population that could look out on a land base apparently damaged by human activities.

The 1930s manifested an environment under stress, most publicized in terms of the Dust Bowl of the great plains of North America but also experienced in less extreme form in other areas. Ontario, for example, experienced crises in the mid-1930s: in summers it was drought, including the drying up of wells and stream courses, whereas at other seasons, ironically, it was flooding. The drought damage was largely in farming country, whereas flooding was an urban more than a rural disaster. Eventually it was recognized that the prime cause of these extremes was the transformation of the terrain from woodland to farmland, from a heavily canopied to a nearly treeless surface, by a century of settlement. In any case, pressure for scientific study of conditions and the design of solutions came mainly from agriculture- and nature-oriented organizations. The conditions were severe enough that two major conferences were convened primarily to devise solutions to head off such crises in future. The report of the first of the conferences, the Guelph Conference on Conservation and Post-War Rehabilitation, apparently was written by the academic botanist, A.F. Coventry. A version of the new ecosystem concept, threading through the report, said "Natural resources form a complex, delicately balanced system, of which each component is sensitive to alterations in the others"; and a related warning pointed out "that all the renewable natural resources of the Province are in an unhealthy state. None of these natural resources will restore themselves under present conditions."[31] One of the existing problems was the destruction of farm soil by "the almost ubiquitous occurrence of erosion."[32] The second conference, in London, Ontario, led to the establishment of a Conservation Branch in a government ministry and to substantial efforts in environmental rehabilitation. These developments in Old Ontario, in the first third of the twentieth century, provided a base of knowledge about how agricultural enterprise affects a woodland environment. Theoretically it could cast light on the expansion of farming, for both settlers and officials, in the newly opening woodlands in northern Ontario and elsewhere in boreal Canada.

Although the twentieth century wrestled with the bitter lessons, the education began earlier. George Perkins Marsh was the major precursor for environmental thinkers in North America, certainly a direct forerunner of Rachel Carson.[33] Lowenthal, in his introduction to the Belknap edition of Marsh's *Man and Nature*, includes a telling exchange of correspondence between Marsh and his publisher. The publisher objected to Marsh's emphasis on "Man the Disturber of Nature's Harmonies" and proposed that humans, as part of nature, work in harmony with it; but Marsh responded "No. Nothing is further from my belief, that man is a 'part of nature' or that his action is controlled by the laws of nature."[34] Carson took up the mission eighty years later to show how close the earth was to experiencing a spring season without the age-old optimistic sounds, because of poisoning of the habitats of many species by the effluents of human living. Before Carson wrote, other observers of nature were listing concerns about unexplained oddities and apparent disharmonies in the natural world. In 1910 the Canadian government established the Commission of Conservation, which spent the eleven years of its existence documenting problems or incipient degradation primarily in the natural environment of the country.[35] Kenneth Boulding summed up the informed, concerned opinion in the 1950s with his arresting image of the earth as a spaceship, which is inescapably a closed system.

The warnings were abroad about the deterioration in environmental quality and about the diminishing quantity of resources. But realization of the limits to the extraordinary cornucopia that had been the lot of western Europe and the New World for five centuries was slow in taking shape. The long trek to the current high level of public concern about the environment was given direction by trailblazers like Marsh, and his scattered precursors, but the running out of good farmland shortly after World War I was, for Canadians, probably the first major jolt on the journey.[36] If there were no more farmland, did that mean the end of new agricultural opportunities, the final dying of the famous frontier? The reoccurrence of serious droughts on the arid margin from 1919, followed by farming difficulties in many parts of the boreal margin by the 1930s, suggested an affirmative answer. The ramifications of such an answer could be immediately and deeply understood by a population that had largely originated in a rural way of life and was still sentimentally and economically tied to it.

This book aims to document the last significant expansion of farm settlement in Canada, and the often painful process of discovering that there was a limit to how far traditional farming could go. The focus is on attempts to farm in marginal (some might argue *sub*marginal) conditions in northerly locations that were rather desperately sought out after more accessible land further south was taken up. The kinds of people pursuing the last, elusive remnants of usable land are depicted in chapter 2, along with an analysis of the demographic characteristics of this surging population, aspects of which echo earlier frontiers. The landseekers of the 1910s through the 1930s faced demanding natural conditions in the marginal areas (illustrated in chapter 3) for which, by and large, they had to learn new adaptations or, in some cases, relearn techniques that had been understood by their grandparents on earlier frontiers in the eastern woodlands. Chapter 4 elucidates the desires of agencies and individuals underlying the search for new farmland, and the forms of encouragement by which often unprepared settlers were tempted to try out unopened or abandoned areas. Chapter 5 offers a series of case studies to illustrate the face of failure and, on the other hand, what appear to have been ingredients for success in marginal areas. Chapter 6 reflects on the expansion into the boreal margin during the interwar period, and on the legacy of land use that has transpired. From reviewing the evidence of modest successes and of efforts that failed, some lessons for establishing new settlement are identified. Because of the amount of attention that has been given to the rather dramatic crises in the dry belt – Jones's "Empire of Dust"[37] – the main concern of this book shifts after chapter 3 exclusively to the more attenuated drama of the northern margins from Québec to the Peace River Block in British Columbia. The northern margins were conceptually the direct successor of the faltering dry frontier, but in a quite different environment.

# 2

# New Territory, New People

Unwise choice of men to go on the land is … at the basis of the failures, which, fortunately, are few. Homesteading in the north requires courage and patience, if not a little farming experience …[1]

… les colons … les artisans obscurs … n'ont pas cessé … d'agrandir les horizons de la patrie et de céder … à "cette passion … pour le défrichement."[2]

The frontier experience called for a physically active population, ideally with many future years to put to the task of "défrichement" (clearing) and farm-making. The marginal areas being occupied for farming from the end of the first decade of the twentieth century through the 1930s probably demanded even more physical vigour than had been typically necessary in North America in the previous half-century. Certainly moving into wooded land added a large component of labour to the clearing process over what had been required on the grassland and parkland.

## WHO WERE THEY?

As with earlier frontiers, the people who moved into the marginal areas after 1910 varied in composition from year to year and in destination from place to place. The further east, the more likely that the settlers would be native-born. Québec was the outstanding case of this, where almost every new farm seeker in Lac St Jean Ouest and Abitibi was French

Canadian (see the example of Joseph Laliberté in chapter 5), but even in Nova Scotia and New Brunswick the native-born held a large majority. From Ontario west across the Prairie Provinces to British Columbia, the foreign immigrant, though not numerically predominant, was the icon of the late farming expansion.

In official circles British (or French) immigrants from "the Mother Country" had been seen as the natural first choice, although ironically the British in the late 1920s and early 30s had roughly three times the deportation rate of other immigrants for being a public charge or a criminal.[3] Public officials apparently categorized immigrants in practice as "preferred" (British, French, American, and northwestern Europeans), "non-preferred" (eastern and southern Europeans), or "others" (excluded), but significant numbers of the non-preferred began to be admitted because they typically had the desired agricultural experience and commitment that became less common among the preferred.[4] Even native-born rural young people had begun choosing a job in town over a fringe farm. Both the pro- and anti-foreign immigration camps recognized that because the traditional British influx had steadily decreased from before World War I, the foreign immigrants formed a relatively larger contingent (though smaller than the numerous expressions of concern at the time might suggest). This demographic drift caught the attention of the census of 1941, in an introductory chapter entitled "Birthplace and the Immigrant Population," where the changes over the previous four decades of immigration are encapsulated:

In 1901, three out of five immigrants were from British countries and two out of five were from foreign countries; now the ratio is about half and half. In 1901, United States-born residents of Canada slightly outnumbered Continental Europeans; in 1941 Continental Europeans outnumbered United States born by two to one. At the turn of the century, only a slight disparity existed between the proportion of immigrants from North Western and from South,

Eastern, and Central Europe; in 1941 the latter outnumbered the former by more than three to one.[5]

In the marginal settlement after 1910 the Canadian-born majority provided the base, although here and there immigrants formed homogeneous demographic and cultural nodes. Some students of twentieth-century rural settlement have described populous immigrant clusters as *blocs* or colonies, including especially those of the Mennonites, Doukhobors, Hutterites, Mormons, German Catholics, and French Canadians in the western interior.[6] Further east there were examples of group settlement but, except for the Québécois expansion into northern Ontario, the numbers were smaller. The British arrivals were both declining in numbers and heading more and more for cities, whereas the central and eastern European newcomers were largely rural people looking for farms. In general, the immigration to Québec and Ontario was predominantly to urban locations. In Québec the skewing was most pronounced, with only 38,810 of 272,146 immigrants living in rural areas in 1941. In Ontario less than a third of immigrants were rural: 221,445 of a total of 782,544. The immigration to the western interior had been much more rural: in Manitoba nearly half the immigrants were rural (93,374 of 198,111); in Alberta well over half (162,014 of 264,708); and Saskatchewan, in keeping with its image, had the largest rural proportion (159,696 of 243,718).[7] The census of 1941 viewed immigration as at the end of an era, pointing to the "virtual cessation of immigration during the 1930s."[8]

Viewing in greater detail the boreal margin midway between the two world wars, in the census of 1931, the difference in the relative size and composition of the immigrant component between Québec and the West was striking. The people moving to the expanding settlements in the boreal forest of Québec were almost entirely born in the province. The largest number of residents born elsewhere were not from inside Canada or any of the common overseas source areas but from the United States, many of them likely of French Canadian extraction. Lac St Jean and Abitibi were quite similar in this respect, although Abitibi, with half the population, had roughly twice as many immigrants: 484 residents born in the United States and 247 in Europe (in its population of 23,692 in 1931), compared to 292 and 173 respectively in the combined Lac St Jean east and west. These immigrant numbers were insignificant among the tens of thousands of the majority population.

The Cochrane District in Ontario, embracing the Great Clay Belt, was much more like the western census districts in having a sizable and mixed immigrant population. In 1931, 9,887 of its 58,033 residents were immigrants from Europe, 4,502 were British, and 1,253 were from the United States. The European group hailed from many different countries, including 2,402 from Finland, 1,669 from Poland, 1,084 from Italy, and 990 from Yugoslavia.[9] A large part of the immigrant population was engaged not in farming but in an industrial activity such as railway maintenance, mining, or forestry, and many of those who settled on farms were also part-time forestry workers. The enticement of cutting trees for paper pulp, instead of focussing on farm improvement, was a particular irritant for the superintendent of the experimental farm at Kapuskasing: "We got a type of settler who was a timber farmer and only took land for the timber, and when it was all cut off moved on to where he could get another lot with a thousand or two cords of pulp on same, till the Department put a stop to this, and the price of pulp has got so low that … they are gradually being driven to the soil where a living can be made if proper methods are followed." (More recent commentators point out that cutting trees made survival sense in marginal conditions, especially when part of the clearing was of one's own land.)[10]

The mixture of the population did not produce harmony in northern Ontario with, as in most parts of the country, its strong whiff of xenophobia and the prevalence of terms such as "bohunk." Arthur Lower tries to put a positive face on his description of the relations between the ethnic groups, saying "Each of these peoples have their qualities. The English-Canadian settlers are said to be, as would be expected, the most progressive and ambitious, but they are followed closely by the Finns and Scandinavians. Nowhere is the English sense of race, always sharp, more keen than among the simple frontiersmen ... It will only be with the utmost reluctance that the English will consent to consider the others their equals and even if the Europeans are converted superficially into English-Canadians, there will always be the French, between whom and the English little real community of feeling or of action does or can exist."[11]

West of Ontario the immigrant component of the population was markedly larger by the 1930s, and showed interesting differences in the source areas of immigrants.[12] In Cochrane District the British-born numbered close to half as many as Europeans and four times as many as United States-born, whereas in the boreal districts of the West the British proportion was generally lower, except where there were important soldier settlements or initiatives under the British Family Settlement scheme. The boreal fringe in Manitoba presented a transition in the immigrant composition, having comparatively large numbers of British in places (notably around Swan River), but in Saskatchewan and Alberta the northern settlements tended to have roughly the same or more United States-born as British, while the European-born in combination outnumbered both. The most important European source area was almost always Poland, although many under that label (as well as some from other eastern European states) would have considered themselves Ukrainian. Across the northern fringe of Saskatchewan the next most important European birthplace of immigrants was Russia, although in the eastern Census Division 14 it was Norway. In the boreal forest areas of Alberta the next most important source after Poland was Romania, except for the Peace River country where it was Norway. The numerical impact of the foreign immigrants in the West is revealed by calculations that show an absolute predominance of non-English speakers over the combined British and American immigrants in half the census divisions of northern Saskatchewan and northern Alberta.

The foreign-language immigrants were not always judged negatively by observers. Some of the testimony to the Saskatchewan Royal Commission on Immigration and Settlement in 1930 revealed grudging respect for the eastern Europeans, primarily because they were willing to endure very primitive living conditions for a long period. The secretary of the White Fox United Farmers of Canada branch, north of the Saskatchewan River and Nipawin in the east of the province, described the scarcity of decent farmland and the extreme difficulty of homesteading in most of the surrounding area. His school district had eighty-seven quarter sections for homesteading, but only eleven of the original settlers were still there after a decade. At this point eastern Europeans were coming to take up land, as he testifies:

"Q. What are the chief nationalities of the people coming in here the last two years?
A. Chiefly Russians and Ukrainians.
Q. How are they getting on?
A. They are existing. They don't live. They have little mud shacks, a few potatoes, a little sugar, coffee, flour and Moose meat. They are trying to make a home and looking forward to something."

And later, in a discussion of the high rate of failure among the original Canadian homesteaders, the White Fox spokesman is asked:

"Q. Of the homesteads taken in 1929, home many, do you think, will stick?
A. I figure about fifty percent will stick.
Q. Why?
A. Because they are a different nationality and they are forming a colony. They are mostly Ukrainians and they will be satisfied with a lower standard of living. They are coming from practically intolerable conditions in their own country. The standard here will be a little better."[13]

A story similar to this was repeated at many places across the northern tier of townships in Saskatchewan and Alberta, into the Peace River country. It usually included wonder at the willingness of the eastern Europeans to endure privations, and a belief that if anyone could make something of the conditions, they could (see more evidence in the discussion of group settlement in chapter 5).

### *And the Next Question ...*

The question has been asked in chapter one: Why would people choose what was widely seen as a fateful gamble in settling on the northern margins? The evidence is not forthcoming or clear, but the intimations are strong. The move to the margins, whether or not recognized by participants, was related to self-worth: for men, the ability to conquer natural conditions and support dependants; for women, the ability to manage a homestead and stand strong for the family. Cecilia Danysk draws attention to the importance of manliness in the settling of the Prairie Provinces, and the intimate connection between farming success and the "construction of masculinity."[14] Almost to the same extent, women in the settler household were held to expectations of domesticity and support of the enterprise. The issue of the "gendered identities of Prairie homestead women" is illustrated by Langford with the reminiscences of a variety of female settlers, including a number in the boreal forest (see especially chapter 5).[15]

We can discover intimations of the pressures on individuals in playing out their roles. A letter from Mrs Peter Stenhouse, writing without her husband's knowledge to the Saskatchewan Minister of Agriculture in April 1936, says "I am in very great trouble ... we have gone steadily Backwards till we have nothing left outside or in that is not Wornout ... if we can't get ... seed it means ruin for us ... and two more Boys to swell the Ranks of the Rod Riders ... my Husband is a good farmer. In all the years we have been farming I have never known him to be as worried as he is now. It looks like the end of everything to him."[16] One can read despair at failing to be the male breadwinner and provider for the family. And a comparable example is seen in a farmer who had been very involved in his community, who wrote to Premier Aberhart from the boreal fringe in Alberta, also in 1936, asking for help with his "financial mess … I would like to know if there would be any chance of getting the debt down to where I could pay and provide a livelihood, or else throw my hands up and take a chance on being able to provide for my family otherwise … I understand that the government is giving assistance in needy cases to farmers to buy feed for stock … I would also like to know if I could get assistance to get seed again."[17] The five crowded pages of his handwritten letter are full of desperation.

Women were expected to play their part. Bearing and caring for children was a classic contribution of the female settler. For one of the wives in the Porcupine Plain settlement, east of Prince Albert, Saskatchewan, her first trip out was to have her first child: "on the trip out the buggy pole broke, the roads were rough … When it got too rough for me, I had to walk … and tread from log to log to keep out of the water too. Every now and again he'd stop and he'd call, 'Are you all right?' I guess maybe I was crying a little. And I'd say, 'Oh yes, keep a-going' … We got to Prairie River. I wasn't fit to be seen … I had to go to Prince Albert by train … I was all alone in Prince Albert … Then I lost the baby, she only lived forty-eight hours. So that was a poor trip … We had to put up with a lot of things here when we first came. Nobody knew very much. So, one woman would go and help another woman."[18] The pioneer woman's credo was encapsulated by the president in opening a Home Makers Club meeting in 1937: "The Prairie Mother in these hard times may be down but not out … We must not let our children see us gloomy"; and the rest of the meeting was largely taken up with topics of self-improvement.[19] Seeking and occupying a farm property may have been intertwined with the frontier man's masculinity, as was the success or failure of his tenure. The responsibility on the shoulders of a woman on the late frontier was maintaining a decent homestead and practising what came to be called the Power of Positive Thinking (see chapter 5). Although the *Why* likely had the components described above for the man and the woman of the house, each case of trekking to the boreal frontier would have been unique.

## DEMOGRAPHIC CHARACTERISTICS OF NEW SETTLEMENT

One might expect the average age of a frontier population to be younger than that of a long-settled area, and certainly below that for the country as a whole. Most newly settled areas had the characteristics of a youthful population. In 1931 the death rate for males in Census Division 16, Alberta (Peace River), was 6.3 per thousand, when the average for 220 counties and census divisions across Canada was 9.5 per thousand (see census divisions on figure 2.1).[20] Data for other northern areas show Census Divisions 14 and 17 in Saskatchewan to have had a male death rate per thousand of 7 and 8.2 respectively, Census Division 14 in Alberta (Athabasca area) to have had 7.2, while in Census Division 15 of Manitoba it was 6.7. The death rate for females in these areas was also markedly lower than the national average in 1931. In the eastern boreal margin, however, where mining and a range of urban jobs were more prominent, the distinction in this rate disappears, with Cochrane having a male death rate of 9.9 and Abitibi of 9.1 per thousand.

C.A. Dawson, writing in the early 1930s about a sample of 332 farms in the Peace River country, said that the average age of farm operators in the newest areas of settlement (the "fringe") was some years lower than in the longer-settled areas. The proportion of young and unattached men among the operators was higher than normal, especially in more remote locations, and families in the fringe, usually being relatively new, tended to have few and very young children. His conclusion: "Pioneer regions are where the adults are in their productive prime and in which the numbers of young

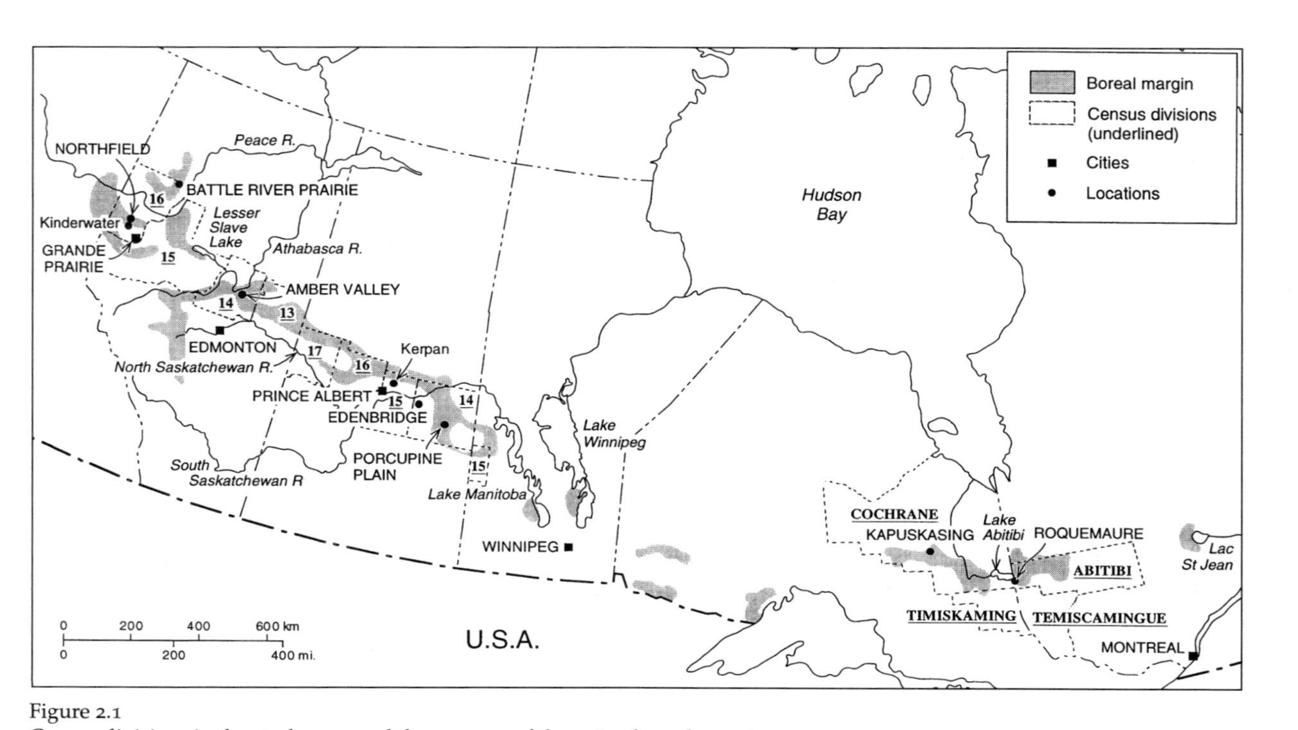


Figure 2.1
Census divisions in the study area and the expanse of the active boreal margin c. 1931.

and aged are disproportionately small."[21] Demographic data, especially from the relatively sparse, striving groups in marginal settlements, can be somewhat mercurial, as Dawson experienced in reassessing the foregoing simple conclusion a few years later. Using data for the same period, Dawson showed that for nine varied sample districts across the Prairie Provinces the average age of farm operators was almost identical, whether in a "stable" or a "new pioneer" area. One consistent result in Dawson's reports was the preponderance of bachelors in the new and "chronic fringe" areas over the stable areas; and another was a strong likelihood of the longer-settled areas having more farmers of age 50 and over.[22] This older age group was more or less balanced by a relatively large number of their offspring reaching adulthood and marrying (thus accounting for the smaller proportion of bachelors). As far as the number of children was concerned, where there was a hint of the farming venture being successful, families responded by increasing the birth rate.[23]

The number of children in the twentieth-century frontier areas usually increased over time, but it did not approach the record of the early nineteenth-century townships in Ontario where it was normal for the population under 16 to account for over 50 per cent of the total during the first fifteen to twenty years.[24] Other nineteenth-century frontiers had similar high numbers of children, usually reaching peak growth about the time that frontier conditions were giving way to moderately successful farming and general improvements in settlement infrastructure. By contrast, only 34 per cent of the population of New Ontario (i.e., the Cochrane District) in 1931 was under age 15. This proportion was repeated throughout the boreal frontier in the Prairie Provinces: e.g., 34–36 per cent in northwest Manitoba; 36–38 per cent in northern Saskatchewan; 32 per cent approaching the Alberta piedmont west of Red Deer; and 35 per cent in the Peace River country. The northern settlements in Québec had somewhat higher numbers under age 15: Abitibi reported 44 per cent and Lac St Jean 47 per cent in 1931,[25] but even these were lower than frontier figures of a century earlier. The dry margin census divisions in Saskatchewan and Alberta had proportions between 33 and 39 per cent.

Frontier populations were strikingly mobile. Not only did the trial-and-error of frontier farming give rise to a great deal of coming and going of hopeful farm families but persons engaging in businesses adjunct to the farming, and others seeking employment as labourers, clerks, and mechanics, were also caught up in the unpredictability and transiency. Even families with a number of children were not immune to moving.[26] In the Vulcan district, a slightly earlier frontier from c. 1904 in southcentral Alberta, fewer than half the farmers who took up land in the early days stayed as long as five years. Gradually land ownership became more general and the economic conditions more stable and after twenty years of settlement, in the 1920s and 30s, close to 90 per cent of the farmers stayed at least five years.[27] Businessmen and employees of all kinds were more transient. In the town of Vulcan, well under half the businessmen of 1920 stayed until 1925, although nearly 60 per cent of those in business in 1925 were continuing in 1930.[28] A great deal of moving in and out occurred during the first years of new settlements, whether in the fertile black soil belt or a challenging late margin. It was a long, repetitive history of transiency on farming frontiers. When farm settlement was beginning in Ontario, in the 1790s and early nineteenth century, a township often would lose half its population in a decade. Transiency was the order of the day as ambitious young farmers moved on, seeking the golden opportunity. Usually a large proportion of the first settlers, having had the choice of the best land, would put down roots in the township. They and their offspring would reinforce their position in the township economy and administration, while large numbers of later land-seekers would reconnoitre the opportunities, perhaps stay a few years, and then move on to greener fields.[29] The transiency continued for decades. Even after the middle of the nineteenth century, townships in Peel County, Ontario, that had been open for settlement for forty years, continued to have a large exodus of population.[30] During at least the first three decades of settlement on an agricultural frontier, the society and its infrastructure were very much "in the making."

In some respects the dry margin differed from the boreal margin in demographic details. The dry fringes in Saskatchewan and Alberta displayed an odd population trajectory between 1910 and 1941. Most of the townships began settlement around 1910 with a burst that continued into the early 1920s, but was followed by a gradual and continuous loss of population (figure 2.2). A local historian visualized it as "a 'see-saw' of economical 'ups and downs.' … when drought persisted, crops were poor, farmers were broke and businesses of the village closed down and moved away. All the grain elevators but one were torn down."[31] That was the situation in the early 1920s in southeastern Alberta. The dry townships were seen to offer very little security for grain farming, and they were almost totally avoided in the 1930s for the redeployment of urban poor to rural land, whereas the northern fringes were seen as a refuge. The repetition of strikingly similar population graphs for sub-humid Saskatchewan and Alberta, seen in figure 2.2, suggests a strong similarity in climatic conditions and social responses.

The northern late frontier was another matter. The colder areas, unlike the dry, were outside the prairie, and were almost entirely beyond the parkland belt into the boreal forest. They often were endowed with more heterogeneous land conditions, and dealt with somewhat variable climatic features under a temperature regime generally lower than that previously acceptable for crop farming. The northern forest areas that were thought to be eligible for farming, stretched not only across Alberta, Saskatchewan, and Manitoba but also into parts of British Columbia and discontinuously across northern Ontario, Québec, and in scattered patches further east.

The Québec government had been zealously promoting colonization for years, and the Ontario and other eastern

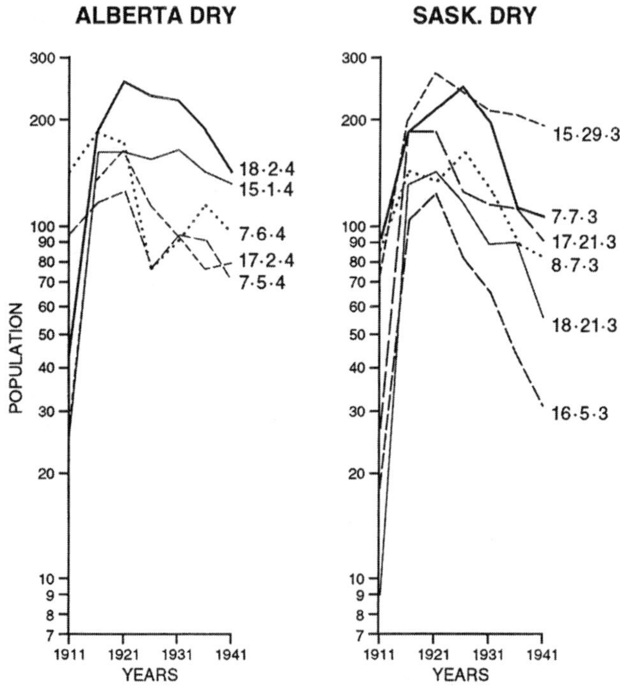


Figure 2.2
Population change in townships in the dry margin of southwest Saskatchewan and southeast Alberta, 1911–41. The pattern in the dry townships is quite consistent, unlike the more variable records in the boreal townships. All the townships and parishes in figures 2.2 to 2.7 were settled after the turn of the twentieth century, most after 1906. Explanation of the numbering of western townships can be found in appendix A; principal meridians are on map fig. 1.2.
*Sources:* Canada *Census 1941*, vol. 2, table 11; *Census of the Prairie Provinces 1936*, vol. 1, table 6.

provincial governments periodically showed interest, with peaks in soldier settlement after World War I and in the Back-to-the-Land movement early in the Great Depression.[32] The federal government and certain of its ministries also had a perennial interest in expanding the country's agricultural territory. The Laurier government, with Clifford Sifton as its settlement entrepreneur, vigorously pursued the opening of the western interior at the turn of the twentieth century. But other areas across the country were thought to show promise, as suggested by the Commission of Conservation's early attention to the so-called "New Ontario," comprising the Great Clay Belt (later included primarily in Cochrane District).[33] The resulting report by the forester B.E. Fernow reveals the highly qualified author to be far from impressed by the agricultural potential apparent from the rail line (see illustration 3.1 and figure 5.2). Of course, in addition to land conditions, the climatic conditions also made agricultural success questionable. But settlement began in many of the townships about this time and went on for at least a couple of decades.

The federal census in eastern Canada was taken once a decade (in contrast to the five-year reporting for the Prairie Provinces, including 1916, 1926, and 1936, in addition to the usual decadal figures). Thus our view of population change in the marginal areas of Ontario and Québec is not as fully informed as for the West. It is possible to reveal the population trajectory on a ten-year basis, for 1911, 1921, 1931, and 1941, and figures 2.3 and 2.4 show graphs for a number of townships in northern Ontario and Québec. A repetitive characteristic of the northern settlements in Ontario and Québec, from c. 1911, was an apparent hesitation in the growth of the population in the first fifteen to twenty years. In a few townships where growth was rapid at the outset, the 1930s brought a loss of population, notably in Ontario, in spite of the attempts by government to relieve the depressed economic conditions by encouraging people in difficulty to move to unused or potential farmland. Although the municipalities in Abitibi and Lac St Jean generally showed unevenness in growth, only one of the examples experienced outright loss prior to World War II.

The boreal margins in the Prairie Provinces commonly had experienced the beginnings of settlement by the second decade of the twentieth century, the date of first arrivals being related to the proximity of the major routes of entry that fanned out from Winnipeg, especially toward Regina and Calgary, and toward Saskatoon and Edmonton. Some

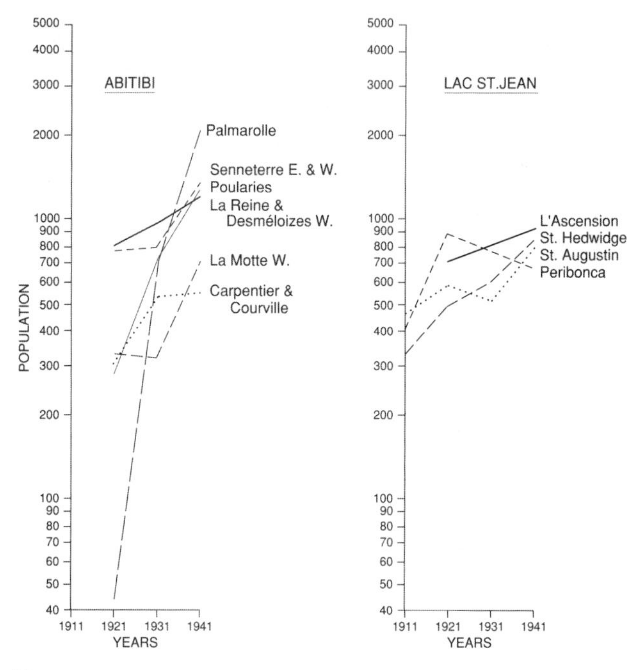


Figure 2.3
Population change in boreal margin parishes in Québec (Abitibi and Lac St Jean Ouest districts), 1911–41.
*Sources:* See fig. 2.2.

boreal townships in Manitoba were settling by 1906, whereas few in Saskatchewan or Alberta had settlers before World War I, except for those on the skirt of the Rocky Mountain foothills between Edmonton and Calgary, and in a forested arc 100 km northeast of Edmonton. In the early years, the small populations in these townships beyond the pale of acceptability could fluctuate widely (as shown on the graphs; the characteristic numerical dominance of males on the frontier is illustrated in appendix A).

In Manitoba, the townships at the northern edge of potential agricultural land experienced a large initial influx around the time of World War I often followed, between Lake Winnipeg and Lake Manitoba, by a steady reduction

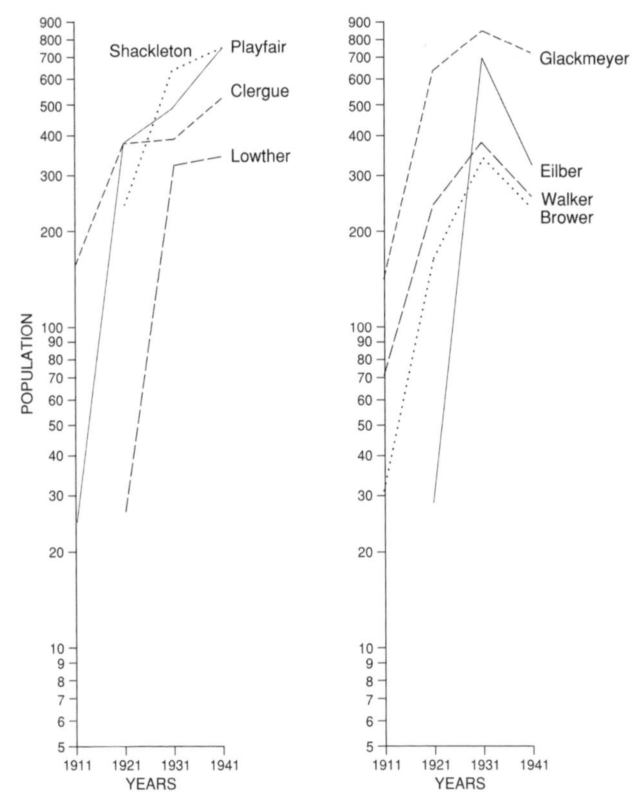


Figure 2.4
Population change in boreal margin townships in Cochrane District (the Great Clay Belt), Ontario, 1911–41.
*Sources:* See fig. 2.2.

through the 1920s with little recovery during the 1930s. The townships near the northwest boundary of Manitoba (in the vicinity of the Swan River district) had a slowly growing or fluctuating population during the first three decades of the twentieth century, but most were at least holding on to their populations by the end of the 1930s. The graphs for the townships west of Red Deer and northeast of Edmonton in Alberta show fluctuations similar to those of the contemporary settling in Manitoba. The only northern townships in the Prairie Provinces that displayed the kind of explosive, "typical" frontier growth from the start, and until at least the late 1930s, were the relatively late settlements north and northeast of Prince Albert, Saskatchewan, and those on the good land discovered on the north side of the Peace River basin in Alberta and British Columbia. Relevant townships shown on the graph include 51.20.2 (near Choiceland, SK), 52.14.2 (near Nipawin, SK), 52.24.2 (near Meath Park, SK), 85.5.6 and 86.5.6 (around Eureka River, AB) (see figures 2.6 and 2.7). A closely comparable Manitoba example is 38.25.1 a part of Minitonas district 20 km northeast of Swan River town (see figure 2.5). These, with the Québec parishes in the eastern boreal forest, supported by the government and the church, were the few exceptions of growth from opening until World War II. A general picture begins to emerge: the demographic experience typical of the marginal settlements was of irregular growth from the beginning, while new land was being tested, followed by what appears generally to have been a permanent loss of population beginning before 1941. This was not unique to the marginal townships – many farming areas, even on the best prairie, were losing people at this time; but it was more widespread and chronic where success was elusive (see figures 2.5, 2.6, 2.7).

## THE MEANING OF THE GENDER RATIO

One demographic measure on which all frontiers agreed was a clear numerical predominance of men over women. Many of the dynamics of the typical frontier were tied to the surplus of men, and perhaps even the definition of frontier could be seen to be an expression of the imbalance: when the number of men to 100 women ranged above 138 those on the agricultural frontier were embroiled in basic clearing and subsistence activities, but when it slipped below that threshold the harsh frontier conditions were beginning to be replaced by a simple social infrastructure, an all-season connection with markets, rudimentary commerce, the appearance of tiny urban clusters (e.g., sawmill,

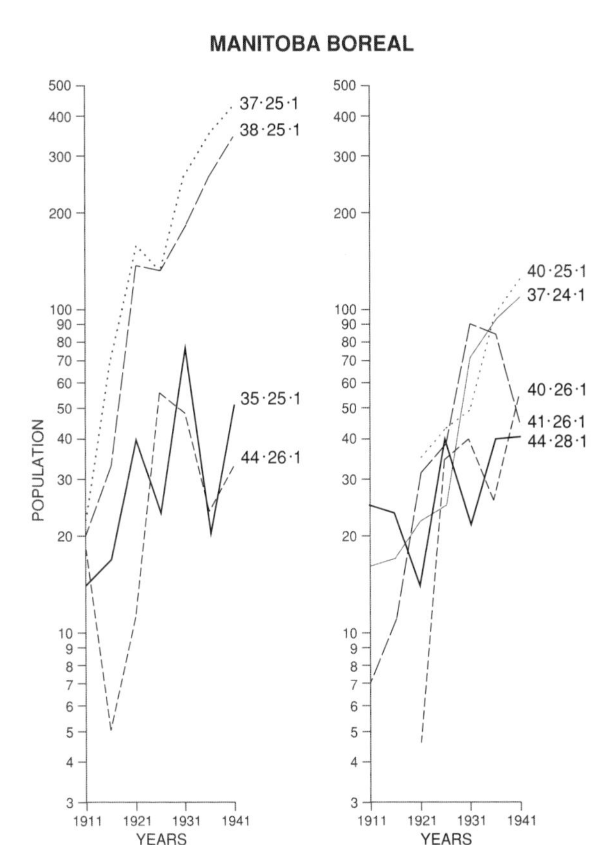


Figure 2.5
Population change in boreal margin townships in Manitoba (Census Division 15), 1911–41. Explanation of the numbering of western townships can be found in appendix A; principal meridians are on map fig. 1.2.
*Sources:* See fig. 2.2.

blacksmith, store), and a state in which all usable farmland is occupied (called "saturation" by Bouchard[34]). A demographic law seems to have applied to the settling of new land, in which the modulating dominance of men went hand in hand with changes on the ground. By comparison, the total male to female ratio for Canada, influenced by long-settled and well-developed parts of the country, was 106:100 in 1921 and 107:100 in 1931.

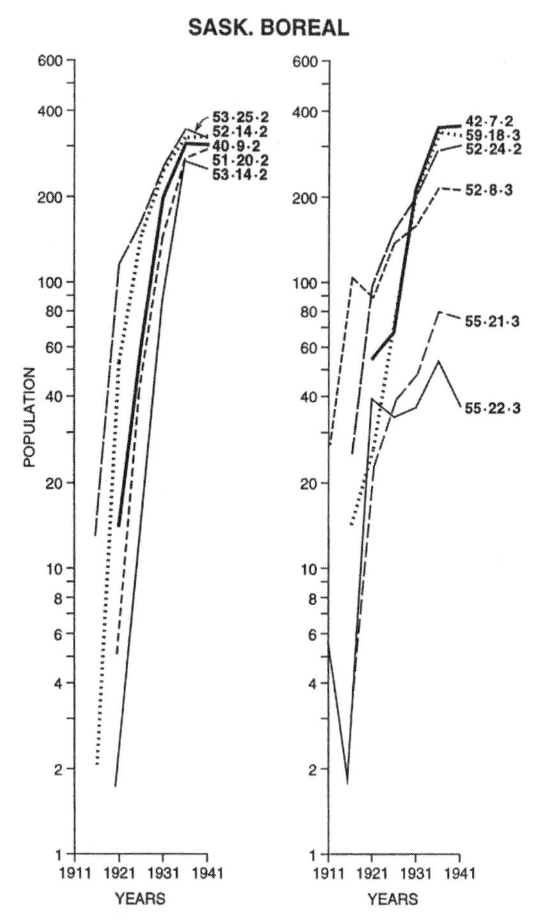


Figure 2.6
Population change in boreal margin townships in Saskatchewan (Census Divisions 14, 15, 16, 17), 1911–41.
*Sources:* See fig. 2.2.

In the first five years of a new agricultural settlement men usually outnumbered women by roughly two to one. This was a natural aspect of the frontier process in which the men, whether married or single, would go in search of a desirable location. Wives and children would follow when a place was found and some preparation had been made (see census figures for males and females in sample townships

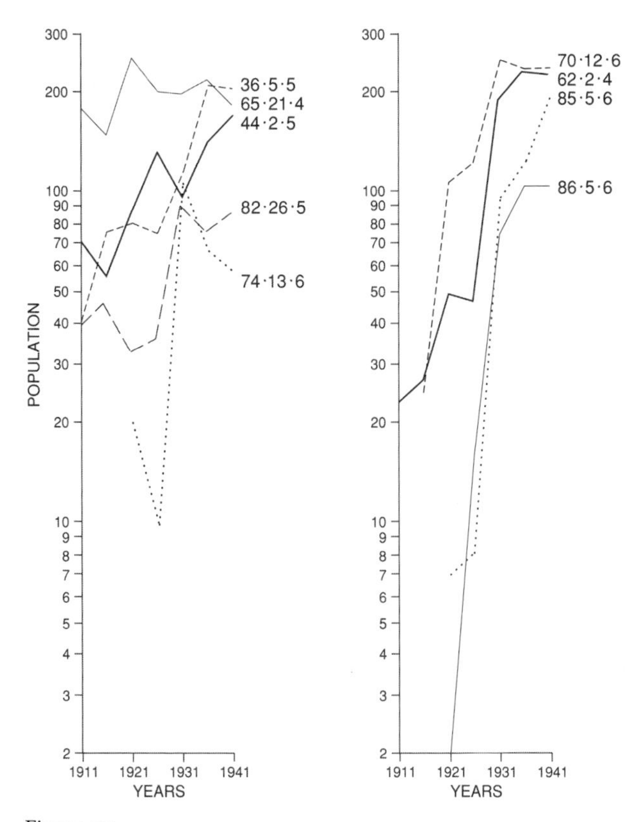


Figure 2.7
Population change in boreal margin townships in Alberta (Census Divisions 9, 13, 14, 16), 1911–41.
*Sources:* See fig. 2.2.

in appendix A). Certainly in a woodland frontier the first few years were notoriously rough, dangerous, and dirty, when trees were being felled, dragged, and burned. In botanist John Muir's colourful word-picture of the "hard-working, hard-drinking, stolid Canadians" that he came across in the 1860s near Georgian Bay in Ontario: "So many acres chopped is their motto, so they grub away amid the smoke of magnificent forest trees, black as demons and material as the soil they move upon."[35] Living conditions were usually primitive and temporary, and the food supply – not to mention schooling and medical services – was very limited. It was not a good place for women and children.

On the eve of World War I the agricultural frontier in Ontario had moved five hundred kilometres north of Georgian Bay and, although fifty years had passed since Muir's observation, the settling process was very similar in the "New" Ontario. Because the reporting of most data is based on a census division rather than on the township subdivision, changes in some characteristics of the fringe areas are somewhat hidden by inclusion in the more general conditions. Similarly, often only total male and female figures are available rather than the more revealing figures, in frontier settlement conditions, for *adult* males and females. As indicated earlier, a mid-decade census was not available for Ontario and Québec, so the view is restricted to the regular ten-year census from the beginnings of settlement just before 1910. The Great Clay Belt, which became the core of the new Cochrane District, had a rural male to female ratio of 165:100 in 1921, and 169:100 in 1931. Townships could vary considerably in their demographic composition, but none of the Great Clay Belt townships came close to equality between the genders in the first two decades of settling (see appendix A). By 1941 the rural population in the Cochrane District recorded a gross male to female ratio of 134:100 – suggesting that pioneering still prevailed outside the industrial towns (for census districts and divisions see figure 2.1).[36]

Across the provincial boundary from Cochrane, the clay belt of the Abitibi District of Québec was the basis for farm colonization beginning in the second decade of the twentieth century. Although the approach to settling the new areas was somewhat different, with much more direction and assistance by government and the Catholic church, and greater encouragement of family colonization, the male to female ratio was still similar to that in other parts of the late frontier. In 1921, not long after settlement began, the overall male to female ratio for the rural population in Abitibi was 134:100; in 1931 it was 131:100; and by 1941 it was 124:100 (certainly comparable to Cochrane's 134:100).[37]

The northern farming areas in Manitoba began opening in the first decade of the twentieth century. In 1921 some of the later ones were still beyond the boundary of civic administration, in the "municipality of Minitonas" (twelve townships abutting the Swan River district on the east). Minitonas had a male to female ratio of 136:100 in 1921. One of the Minitonas townships (township 37, range 24 west of the prime meridian) had a ratio of 154:100 in 1931 and 138:100 in 1936 (for clarification of the township-range survey, see appendix A). What most of the townships demonstrate is one of the common features of frontier evolution: the longer a new agricultural settlement persists, the closer the gender ratio moves toward equality (not to deny that aberrations do occur, as revealed in appendix A). The demographic mechanism at work is primarily the establishment of families and the increase of children.

Reviewing the data for the northern fringe settlements in Saskatchewan reveals that in the 1930s the male to female ratios generally hovered around 130 to 150 males per 100 females. Almost all the boreal forest margin townships in Saskatchewan were settled at the end of World War I, and after fifteen years of pioneering activity, in the early 1930s, many were still attracting large influxes of males. Townships in the vicinity of Porcupine Plain, part of Census Division 14, had about 129 males for 100 females in 1931, reducing to about 118:100 in 1936 (see the review of the Porcupine Plain Soldier Settlement scheme in chapter 5). The boreal forest frontier in Alberta showed a variety of demographic conditions, but the gradual reduction in male numerical dominance with the passage of time also occurred there. Conditions in the new farming areas were best represented by the figure for *rural adults,* where available. By that calculation the ratio for the forested margin west of Red Deer, that began recording population in 1906, was 151:100 in 1936.[38] By comparison, in Division 14, northeast of Edmonton, the contrast between numbers of males and females was generally more muted in the 1930s in townships that had begun settlement just before 1911. For the division's rural adult population the ratio was 138:100

in 1936. The Peace River country, Alberta (Census Division 16), was close to being a classic frontier area, with rapid continuous growth from the start of farm settlement. New land was still being opened in the 1930s in some outlying parkland areas (as it was in adjacent British Columbia). Census Division 16 illustrates the modulation in the gender ratio: the census of 1911, the first after farm settlement began in 1907, indicated 213 males for 100 females; in 1916, 208; in 1921, 158; and in 1926, 137.

While the census division and even the provincial ratios of males to females were above parity, those of the marginal settlements were invariably higher, ranging around 130 to 190 males per 100 females throughout the first two decades of settlement. Where data are available for rural adults, it is clear that the major reason for the imbalance is related to the kind of rough work involved in clearing forest to make farmland. The inclusion of children, as in most of the figures in appendix A, tends to pull the ratio towards parity because boys and girls are born in roughly equal numbers. Similarly, the inclusion of urban population tends to pull the ratio towards parity because towns usually attracted similar numbers of males and females, and not uncommonly had a surplus of working-age women. An additional gloss may be put on comparisons between some of the Québec parishes and many townships in Cochrane District and the West: after only ten years of settlement, in 1931, La Motte W. and Roquemaure parishes were approaching equal numbers of males and females, which suggests the success of the province's colonization efforts based on families. Similar success was not, however, being shown by all the district's parishes (cf Senneterre, and Poularies).

## ABERRATIONS AND REGULARITIES

Frontiers have a reputation for attracting unusual individuals, many of whom appear to be "putting a past behind them" (although this is probably no more common than in an inner city neighbourhood). The outer edges of settlement also could be attractive to groups out of step with the society in general, in the sense of being nonconformist. Thus the Hutterites came into western Canada, beginning in 1918, searching for some isolation for their communal farming.[39] The Doukhobors sought a refuge in northern Saskatchewan for their distinctive way of life; and even conservative Jewish groups located for a time on the wide-spaced quarter-section farms of the prairies.[40] Mennonites too looked to some marginal areas as possibilities for expansion. This colourful patchwork of ethnic groups, some of which had been persecuted or otherwise marginalized in their areas of origin, was a focus of the attention of the Canadian Pioneer Problems Committee in the early 1930s. The work of this committee gave rise to a remarkable map depicting the kaleidoscopic pattern of ethnic clusters across the Prairie Provinces, and about the same time observers were writing with incredulity about the odd habits and lifestyles of immigrant groups.[41] It seems that the first recognition in Canada (excepting perhaps the confrontation with the First Nations) of what is today called The Other came from pictures of the poverty and grime of bewildered immigrants from old cultures in Europe being thrust in the face of prim Canadian society. This was a bone to be gnawed by the eugenicists and racists around the time of World War I, and it provided fodder for the debate over immigration policy.

"Otherness" was certainly a characteristic of many of the groups on first arrival on the frontier, even of some that eventually blended into the establishment, such as the religiously nonconformist "Bull Outfit" (predominantly Burnsites from southern Ontario, characterized by a fundamentalist emphasis on holiness[42]) that led the entry to the Beaverlodge area of the south Peace River country. Rural areas, especially partially developed fringes, could shelter "otherness" for a time – as they do even today – because of the easy avoidance of social contacts and the availability of rustic hideaways.[43] Often it was the "morality" of the outlying groups that was claimed to be a concern. In his study of the eastern forest frontier in the 1930s, Arthur Lower included as an appendix a questionnaire from northern Ontario asking teachers to report on the "qualities" of their tributary population, in which assessment of "racial" groups loomed large, and on the home life of the average pioneer: "Dull, depraved, decent, character forming, ambitious?" (see groups discussed, chapter 5).[44]

Notwithstanding the vaunted variety and nonconformity particularly of the late settlement in marginal conditions, it is possible to identify common characteristics (or *regularities*) in the way frontier settlement was played out. Some examples, as elaborated above, include the proportion of children and the male to female ratio. Gérard Bouchard proposes a more general model that he argues can embrace such disparate expressions of farm settlement as Québec's Saguenay, anglophone eastern Ontario, and extensive parts of northeastern and northcentral United States. The central feature of Bouchard's model is what he claims to be the primary goal of farm settlement, namely to sustain and progressively enhance the farm family's success, or in a broad sense "family reproduction."[45] His model was replicated in many New World frontier settlements under conditions of an "open system," in which there was an abundance of usable and alienable land, a rapid increase in the number of children, an expectation that each son would be provided with livelihood, and the growth of egalitarian social relations. The conditions fostering such an open system, however, had been available only before World War I, with few exceptions, some of which appeared on the boreal margin. But whatever the apparent common social and economic structures that frontiers seemed to generate, the late expansion toward the north in Canada came to be characterized as much by *closing* as by *opening* of opportunities. The Ontario government opted for closing its farm settlement efforts in the northern reaches of the province after nearly three decades. As Premier Hepburn declared in April 1935, "We are going out of the business of colonization ... If farmers in southern Ontario, on good land and with all facilities, are unable to get along, what chances have new farmers on barren land in the north?"[46]

The limitations faced by frontier people were illustrated copiously by those engaging in or observing the incursions into new land after 1910. Across the country, the expansion of agriculture met climatic restrictions. The move toward the north had to deal with the reality of generally colder conditions (as detailed in chapter 3). The newspaper *La Terre de Chez Nous*, organ of L'Union Catholique des Cultivateurs, which encouraged new farm settlement, was at pains in the 1930s to reduce concern about the coldness of the northern woods. In an article entitled "le Climat Abitibien," it said that the inability to have a harvest in the first few years because of cold "is somewhat the case in our northern areas ... as long as they are not sufficiently cleared and drained. For the land to warm up, it is necessary for the sun's rays to penetrate it. And in the areas of coniferous forest ... the mosses that cover the soil hold the frost and ice till a late date, into summer ... In a new country, like Abitibi ... to the extent that clearing is done the climate is ameliorated."[47] The expectation was that extensive clearing would result in more favourable conditions for agriculture, perhaps even comparable to southern Manitoba or better parts of Québec. The authorities who fostered the expansion counted on the settler being strongly abetted by a wife, and the spasmodic verbal encouragement of the distaff side was florid and unctuous: "the colonist ... was not alone. His pioneering enthusiasm was supported by the quiet generosity of the well-run home base, managed, inspired by a wife ... she also shares the conviction that soon ... they will cut a tree that will be the last, that 'the land will be made,' and that they will undertake ... the last stage of a life of labour and long-suffering."[48]

Many people struggled through the difficulties to various degrees of success. One example of dealing with the problem of poor access in the Great Clay Belt of Ontario was reported in the local paper: approximately fifteen miles north of Cochrane, a Mr Bentley had grown some creditable produce despite "the most discouraging conditions as far as communication with the outside is concerned. The last three or four miles are leading over what is called a road, so atrocious as to make the passing a hard task even under the most favorable weather conditions. Mr Bentley is not the only settler up there, and years ago solemn promises were made."[49]

In the northwestern end of the boreal margin, in Alberta north of Edmonton, evidence shows that even persons who had been influential in their community were not always immune to the scourge of farming failure. Writing from the farm he had worked for twenty-four years, T.D. Cunningham continues his plea to Premier Aberhart: "farming is all I have done all my life … I … have been trustee for our school board for 23 years … 8 yrs as chairman … and was from 1928 to 1934 Sec. treas. U.F.A. [United Farmers of Alberta] Provincial Assn of St Albert. Trusting that you will be able to help me."[50]

In the marginal lands the state of "saturation" or closure of the land boom came rather quickly, usually because parts of each township were obviously not suitable for farming (as the excessive admixture of land quality shown on the Canada Land Inventory maps in chapter 5 attests). Evidence of the limitations to the land budget on the margins appeared regularly, from Fernow's less than positive assessment of New Ontario to Eggleston's wistful farewell to the dried-out homestead to Albright's report on ill-chosen high land in the Peace River country. One sad example came before the Executive Council of the Alberta government in 1934, concerning a destitute settler who had been charged with trapping muskrats without a license: "WHEREAS it appears that part of the Olyniuk farm is in White Rat Lake [northwest of St Paul], and that he only trapped because he was in desperate need of funds to purchase the bare necessities of life for his wife and three small children," the Council ruled that the $10 fine be remitted.[51] The vignettes above provide snapshots of real people trying to prevail on the late frontier. They appear fleetingly in the public record generally because of some substantial misfortune, but in many ways they represent the hardships endured by most of their contemporaries engaged in expanding farming beyond the previous limits of the first twentieth-century decade. Settlers who left a more extended record of their frontier world will be found in the case studies of chapter 5.

The saturation of available land is reflected in the graphs of population change (figures 2.2 to 2.7), where the locus reaches a high point and thereafter levels off or begins to decline. The Bouchard model provides a valuable conceptual bridge to understanding and comparing the basic dynamics of the agricultural frontier in northern North America up to the end of the nineteenth century. But the version of the model that survived into what is called here the late frontier was seriously undernourished: population usually was not undergoing rapid long-term increase, there was little assurance of setting up a number of sons on decent farms, and the striking majority of children in the open system model was greatly reduced on the twentieth-century margins, usually by a quarter to a third. In summation, saturation came quickly in most of the late frontier townships.

In looking back on the peopling of the margins in the first half of the twentieth century, one sees an ebb and flow in what was popularly known about the migrants, their experiences, and the places they were choosing. The period under study showed a continuing agitation for more farmland, primarily by native-born Canadians, and a steady movement, though less celebrated than on earlier frontiers, toward the northern fringes. This went hand in hand with the improvements in transportation, especially the extension of transcontinental and local railway lines, the campaigns to move settlers to newly opened land, and activities like the harvest excursions that only petered out toward the late twenties. Overlying this matrix, however, were movements that caught the attention of the media and gave a certain flavour to different parts of the period. In the 1910s, the tribulations of some of the Doukhobor groups were newsworthy, and the arrival of Hutterites was noted. The last years of World War I and the 1920s were overlain with publicity and debate surrounding the Soldier Settlement schemes across the country. Following on the heels of, and partly intertwined with, the soldier settlement was the British Family (or 3,000 Families) Settlement Scheme. The mid-1920s also saw an influx of late Mennonites from Russia, a small proportion of whom settled in northern Ontario, sixty kilometres west of Kapuskasing. In 1926 a leader reports "we have been here only six months ... We hope to carry out here ... a future with a religious life such as we formerly enjoyed in Russia. The settlement [later called Reesor] numbers about seventy souls," with more families expected from Russia and elsewhere.[52]

In the early 1930s the most widely scrutinized migration to the boreal margin was of welfare cases from the cities, in what was euphemistically called The-Back-to-the-Land Movement. In most provinces this was quickly branded a failure. The so-called "Relief Land Settlement" co-operation between the federal and provincial governments in the 1930s moved on from attempting to rescue welfare recipients to become a massive infusion of mainly federal funds, applied primarily by provincial administrations, to help maintain much more than the poorest stratum of society in the generally depressed conditions. In Québec, and less so in other provinces, the various federal acts were mirrored by provincial acts to generate major infusions of new farm settlement in marginal areas. Perhaps the last large movement related to the late frontier, though in a negative sense, was of the thousands of disillusioned residents of difficult fringe areas who, with the opening of World War II hostilities, grasped the opportunity to leave the farming margins.

# 3

# Old Nature and the Challenges to Farm Settlement

The country is promising ... to the settler ... who is willing to rough it a bit, and to forgo many of the customary comforts ... and who above all is willing for a considerable period to glean his living from his garden and his livestock, for cash income from forest and cut-over farms are inevitably small for some years.[1]

Although the vaunted "free homesteads" of the classic frontier were all claimed in the desirable parts of Canada by the early years of the twentieth century, demand for land to farm was still strong. The only land available for settlers had previously been deemed too remote, or not suitable for crop farming; or it had been damaged and abandoned; in almost all cases it was marginal in farming terms. The Canadian government had publicized the western interior of the country as The Last Best West, a natural successor to the Best West of the United States. But by 1918 settlement had filled The Last Best West, had invaded large stretches of boreal forest in the three Prairie Provinces, Québec, and Ontario, and had even jumped over 120 km of poorly drained forest into the newly opened outlier of the Peace River country of Alberta which, with the adjacent Peace River Block in British Columbia, might be called *The Very Last* Best West.

The frontier idealized in the early twentieth-century expansion was the classic agricultural frontier based on the production of harvested field crops. The primary crop on this frontier, certainly from the Great Lakes basin and across the western interior, had been wheat. There were fundamentally two kinds of late farming frontier: the dry – that played a major role in the sad drama of "the dirty thirties"; and the cold, or more accurately, as shown below, the boreal – almost entirely in the challenging habitat of the forest. These two main categories could be further divided into an *evolving* frontier – an area on the path to some form of permanent land use for farm crops; and an *inherent* frontier – an area with built-in negative characteristics, such as salinity, excessive stoniness, summer frost, or wetness, that would continue to defeat use of the land for food crops.[2]

## FRONTIER AND MARGIN

The frontier has been commonly understood as the outer limit of rural, primarily farm-based, settlement. In F.J. Turner's phrase describing the settlement of the North American continent by Euroamericans, it was "the hither edge" of the territory they occupied. In the Old World the concept of the frontier had been well known – whether in the context of the Welsh Marches to the west of England, or on the lawless borders between England and Scotland, or between the various crystallizing states in mainland Europe.[3] As late as the 1870s a concept of frontier embraced the eastern *départements* of France as "provinces frontières." Forests were a characteristic landscape in this frontier, and at the time the military-dominated Joint Commission on Public Works had jurisdiction even over private forests.[4] In all these cases the frontiers were eventually incorporated into the permanently occupied territory of the adjacent dominant state: these were frontiers able to evolve. From these and other cases had grown an age-old optimism in which a frontier was expected to eventually develop into a useful and productive part of state territory, that is, habitable land or ecumene.[5] Typically the nineteenth-century North American frontier had not been marginal by nature, in terms of the inherent character of the land and the climate.

The positive idea of the frontier as opening the door to new productivity was adopted in North America and loaded with conceptual "baggage" by nationalist writers led by F.J. Turner at the end of the nineteenth century. He argued that the frontier and its attendant opportunities were a basis for the way of life in the United States. But by the 1920s the optimism of the frontier expansion was forced to face physical realities that were not open to amelioration. Many of the so-called frontiers being taken up for settlement at that time were not on the way to becoming good areas for farming: they were beyond certain limits, and would continue to offer only a marginal level of existence as long as farm occupation continued. These inherent frontiers revealed themselves not only at the arid and cold edges of Canada's agricultural land, but at many locations across the country as settlement spread. Some have become areas of permanent farming difficulty or failure. Examples include the thin-soiled peninsula of Nova Scotia, the rolling hills paralleling the boundary between Québec and New England, the light sandy heights of the Oak Ridges Moraine in southcentral Ontario, the inadequately drained and coarse-soiled parts of the Interlake region of Manitoba, and the two hundred kilometres of lake-dotted terrain between Cold Lake and Athabasca in Alberta. Areas like these would come under Dawson and Younge's label of "chronic fringes." The word "fringes," like the word "margin," includes both a declaration of periphery and a hint of low quality. It could be argued that for the chronic fringes or inherent frontiers the term "submarginal" might be nearer the truth.[6]

The northern late frontier being tested by farm settlers was marginal on a number of grounds, including:

1) *remoteness* – being so far away from transportation routes that commercial crop farming was untenable. In the first three decades of the twentieth century the railways were steadily eliminating this particular measure of marginality in all areas where farming was showing promise. A filigree

of lines was laid across the interior plains, even into the wooded areas, and transcontinental lines traversed the boreal forest areas of Ontario and Québec. About the same time, most provincial governments were engaging in road improvements; and during the Great Depression they used road work as a welfare agency.

2) *soil quality* – the soil on the northern, cold frontier in the boreal forest was a poorly endowed grey woodland soil called podzol (from the Russian word for "ashes"). The podzol soil (in more modern terminology, it would be included in the soil order Spodosol or, in the west, Alfisol)[7] was characterized by the leaching of nutrients down through the top layer by precipitation. The leaching left the surface soil deficient in soluble nutrients. The soils in the eastern woodland occupied by settlers during the nineteenth century also were podzols, but there the variety of deciduous trees had provided a deeper accumulation of fertile organic matter than was found under the boreal forest. The northern areas of settlement expansion in Québec and Ontario shared with the northern margin of the Prairie Provinces a variegated land surface and podzolized soils under the typical boreal forest of needleleaf trees with an intermixture of broadleaf species. The soil on the dry frontier, usually described as brown or reddish brown, formed primarily under short-grass prairie. In southern Saskatchewan and Alberta the surface texture could vary from clay and silt or clay loam to light sandy loam and fine sand.[8] The grassland vegetation generally had deep and dense root systems, and a relatively high rate of decay without leaching. Thus organic matter was available in the ground for the first few years of farming. But the lack of regular precipitation and of substantial vegetation as windbreaks, and the using up of the organic matter in the soil, soon rendered the soil vulnerable to wind erosion in particular. Commonly the lighter, more vulnerable soils, being easier to break and till, were the first to be ploughed and cropped, only to succumb to erosion.

3) *land characteristics* – in certain places on both the boreal and the dry margins characteristics other than low soil quality

could hamper or prevent farming. These included poor drainage (tackled in some major, government-sponsored projects after World War II), excessive stoniness in post-glacial beaches, and salinity, especially under primitive irrigation in dry areas. Careless irrigation leading to extended evaporation could raise marine salts and deposit them on the surface as sterile white flats, visible today.

4) *climatic conditions* – the two categories of late frontier are characterized by a dominating climatic feature: on one hand a short growing season (the boreal); and on the other a deficiency of available moisture (the dry). These characteristics had been identified by explorers, government surveyors, and earlier settlers, and this had caused a slowing down of the expansion in certain directions.[9] But climate is variable, and undesirable characteristics can diminish for a time and give rise to revised opinions on the suitability of unoccupied areas for farming. Witness the saga of the famous Palliser's Triangle, the arena of the dry late frontier, the reputation of which went through oscillations for decades from the 1850s to the 1930s (see Palliser's Triangle in figure 1.2). Although less publicly debated, the opening to survey and settlement of the northern fringes was also an example of the periodic repositioning of attitudes about climatic limits to farm settlement. A striking example, regarding the Peace River country, is the divergence of two Macoun opinions: in the 1870s John Macoun (expeditionary colleague of Sandford Fleming) declared the area "fit … to produce anything," whereas in 1904 his son James, perhaps impressed by its distance north, was unimpressed enough with its apparent farming potential to call it "emphatically a poor man's country."[10] A meaningful modern measure is the number of growing degree days (introduced in chapter 1). The depiction of the boreal frontier in figure 2.1 can also be described in terms of growing degree days, that is, in the range of 1250 to 1500 GDD. This compares to sites in older farming areas such as south central Ontario (2200 GDD), Winnipeg (1750 GDD), Saskatoon (1550 GDD), and Edmonton (1500 GDD). A climate with 1350 GDD could ripen spring

wheat, barley, oats, buckwheat, canola, and hardy vegetables, in the "unboreal" expectation of other necessary conditions being optimal.[11]

The foregoing characteristics of the new areas being offered to settlers after the first decade of the twentieth century raises questions about judgments that were made. One might wonder about the accuracy of the available geographical knowledge and also about the motivations that were at work. A question that needs to be pondered is whether or not the variability and unpredictability of the natural conditions (laid out in detail in what follows) and the deficiencies in the knowledge and the mode of decision-making provide justification for the encouragement given to the expansion, especially by government agencies and various organizations.

## MARGINAL CONDITIONS, SETTLEMENT DURESS

Farming problems had already been experienced in the colder parts of central and western Canada. The main agent had been late or early frosts that killed off the wheat crop in tender leaf or just before harvest. Ontario north of the upper Great Lakes and the adjacent portion of Québec, were well known to have a short growing season.[12] In the west, frost had maintained an effective barrier to expansion into areas north of the so-called parkland belt (a huge arch, sometimes called a "fertile crescent," across the three Prairie Provinces, from a northern limit near the North Saskatchewan River, curving southwest toward Calgary, and in the east into southern Manitoba). Active experimentation and cross-pollinating to produce quicker-maturing varieties of wheat were being pursued, and by the end of the first decade of the twentieth century results were promising, although dissemination of seed took time (as Albright found in the Peace River country). During the next two decades, the new wheat was to entice settlers to venture dozens – and in the Peace River country hundreds – of kilometres farther north from the previous line of settlement. The resulting optimism encouraged the belief that it was just a matter of time till another improvement would allow farming to go even farther into hitherto forbidden territory. But some immovable barriers were to materialize.

Frost was the most publicized limitation to agriculture in northern Ontario and Québec, and in the other scattered pockets of settlement expansion further east. Another was a relatively low accumulation of heat over the growing season – a deficiency of growing degree days – because of cloud cover and a lack of hot days, so that little more than hay, root crops, and perhaps oats, could confidently be grown to maturity. A study of four to eight years of colonization in northern New Brunswick in 1939 concluded that the climatic conditions allowed the growing of "ordinary" farm crops: "grains, hay, grass and particularly potatoes. Crops needing warmth, such as corn and soya, do not do as well. The winters are long and cold but the temperature is moderate during the summer."[13] This assessment is compromised, however, by its reliance on climatic data from the experimental station at Fredericton, when most of the Crown land available for colonizing in northern New Brunswick was approximately a thousand feet higher in elevation, and thus less promising for farming.

A post-World War II study of northern Ontario pointed out that "The climate is marginal for many valuable crops and presents difficulties to the farmer in the handling of the soil, crops and livestock ... Farming in the Northern [or Great] Clay Belt is a difficult undertaking ."[14] The difficulty facing farming is illustrated by the acceptance from the beginning of settlement that wheat was not a practical crop. Although it has a mean frost-free period of 92 days, the Great Clay Belt, like the northern New Brunswick upland, has a relatively low build-up of heat for crops, and has relied on hay and other crops that can tolerate a little frost. The mean date of the last spring frost is June 8 and of the first fall frost is September 7.[15] The Mennonite settlement at Reesor got an early indication of the exceptions to the rule in the Great Clay Belt when their first vegetable crop was

Illustration 3.1
Father and son in the midst of clearing boreal forest "in the Clay Belt of northern Ontario" (Commission of Conservation caption).
*Source:* Library and Archives Canada (hereafter LAC), Commission of Conservation collection, PA186746.

hit by frost in 1926: "The summer was short with much rain. In August we had frost that froze all our vegetables."[16] Although it lies at roughly the 49th parallel, the rating of the Great Clay Belt for growing degree days (about 1340, based on 5°C as the start of crop growth) puts it in the same category as the cold margin across Canada, namely the northern Interlake region in Manitoba, the southern edge of the boreal forest in Saskatchewan across to Athabasca, Alberta, and the forest aureole around the Peace River country; and in the opposite direction Québec's Abitibi, much of the Gaspé Peninsula, and northern New Brunswick (see illustrations 3.1 and 3.2). The ominous continental effect in northern Ontario should have been recognized by the early twentieth century because the general form of the east-west climatic belts, and especially the swing toward the northwest, had been fairly well understood by students of climate nearly a century earlier.[17]

Illustration 3.2
Commission of Conservation photograph with the caption "Settler's shack, Ontario. Fine birch, spruce and cedar cut down to make a useless farm."
*Source:* LAC, photographic archives, PA186745.

Along the boreal margin in Manitoba, the official growing season, dated from the time of seeding to the first killing fall frost, was close to 110 days according to a 1920s map by the Natural Resources Intelligence Service of Canada. But at the same time, Dominion meteorological reports showed that the boreal margin could expect a late frost up to June 14 and an early fall frost at the end of August, leaving about 77 "safe" growing days for tender crops. For the frontier as it spread westward, the climate desired was one that would grow wheat. In the nineteenth century in the eastern provinces the chief frontier grain had been some variety of winter wheat that was planted in the autumn and matured the following August. In the Prairie Provinces the grain was spring wheat, normally planted in the first half of May and requiring 90 to 100 days, all frost-free after the emergence of the first leaves, to mature. Most of the eastern Prairie Provinces had on average a long enough frost-free period to mature spring wheat. The exceptions were the higher land on the west side of Manitoba and a large part of the Interlake region north of Gimli. In Saskatchewan a large swath of territory abutting the Manitoba boundary, between Porcupine Plain and Swan River, had an average frost-free season much too short for growing wheat and other tender crops. The pattern of the average frost-free period did not show a simple gradation to a shorter and shorter season toward the north. The picture was quite complex, varying considerably from place to place in response especially to elevation. Surprisingly there was a longer period without frost adjacent to the Saskatchewan River valley northeast of Prince Albert than in a long stretch of the open prairie southward to within 100 km of Regina.[19]

In Alberta, higher areas had a shorter frost-free season, including the western and northwestern edges of the Peace River country and the north-south stretch of the Rocky Mountain piedmont. The whole boreal fringe in Alberta and adjacent Saskatchewan might have a late spring frost up to June 15, based on data from 1951–64, and the area from Athabasca west and north might have a first fall frost in the first two weeks of September (see the more focussed view of the Beaverlodge district below).[20] A review of the policies and process of agricultural settlement in the 1930s in the northern Peace River country of Alberta and British Columbia claimed that, while soils might be surveyed, climatic evaluation was neglected. Only extremely general climatic information was available, because official surveys concentrated on soil quality, terrain, and vegetation. Only much after World War II did research show conditions similar to the Saskatchewan and Manitoba portions of the boreal margin, under which "crop failures of varying degrees ... may be expected one year out of every four, primarily as a result of unseasonable climate (too wet, too dry, or too cold) during a critical stage."[21]

The founder of the agricultural research station at Beaverlodge, W.D. Albright, began keeping weather records during World War I, thus making possible a fuller climatic picture of the western wing of this study. A review of forty-five years of temperature, precipitation, and various related measurements was carried out in the 1960s.[22] Beaverlodge is in the former parkland of the south Peace River area; note that both its average temperature and its precipitation are slightly higher than surrounding areas to the north and west, namely the fringes being settled after World War I. Although the Peace River country had a reputation of being a northern Promised Land, the climatic realities included an average chance of a damaging August frost once in seven years, a common lack of summer heat only partially offset by the long period of daylight, and in spring west winds strong enough to damage sprouting plants and dry out the soil. The average period above freezing, at 101 days, and above killing frost (28°F or –2°C), at 132 days, is more than enough for the ripening of spring wheat; but "In 1 year out of 10 the last frost in the spring comes on June 11 and the first frost in the fall on August 12," while normally the last killing frost in spring is May 8 and the first in autumn is September 20.[23] Only a one-month period, from June 20 to July 20, had been free of frost for all forty-five years of the record, an indication of the dimensions of the challenge of opening northern land to agriculture. In a study of agricultural settlement in the northern Peace River country of Alberta, as far as the Ft Vermilion vicinity at nearly 59° north latitude, Ehlers shows that the further north the colder the winter but, surprisingly, the warmer the summer quarter of the year. The Fairview district north of the Peace River is, in many respects, a homologue of the Beaverlodge district in the south Peace. Ehlers records the mean data of the last spring frost at Fairview as 25 May and the first autumn frost as September 7, a frost-free season of 103 days. The shortest frost-free period was 78 days, much too short for ripening wheat.[24]

The hazard for wheat growing was not just in the average conditions but in the extreme variability of the climate from year to year. Most of the areas occupied for farming in the 1910s, 1920s, and 1930s were near the northern limits of agriculture, so a small deviation from the average could place the wheat crop in jeopardy. Apart from southwest Manitoba and south of Regina in Saskatchewan, the farmland in the eastern Prairie Provinces had a 25 per cent chance of a late spring frost in the first week of June, or an early fall frost around the beginning of September.[25] The level of climatic hazard in the 1920s and 30s was likely higher than indicated by these percentages, because of the muting effect of the length of the record used in the Environment Canada study. The Murchie-Grant study, focussing on the period from the beginning of the century to the 1920s, indicated a considerably later date for the last spring frost and an earlier date for the first fall frost on the northern margins.[26] Also in Saskatchewan, climate needs to be assessed on a decadal rather than a half-century basis: the location and severity of conditions could vary substantially within a ten-year period. In the 1930s, and less markedly in the 1920s, Palliser's Triangle was drier than in any of the later decades in the twentieth century (when, ironically, steppe conditions have been recorded more outside than inside the traditional dry margin).[27]

Enough water at the right time for the crops was a critical climatic factor for farming, though less so in the northern fringes than on the grassland or parkland. East of Manitoba sufficient moisture was almost never a concern; indeed, an excess of water was as likely. The boreal margin in the Prairie Provinces could usually count on having enough moisture to bring a spring grain crop to maturity. For example, within the general scene of almost total failure of the Saskatchewan wheat crop in 1937, a few municipalities north of and east of Prince Albert reported average yields in the range of 15 to 30 bushels per acre.[28] For the growing of crops, moisture had to be accumulated in the ground from winter snow in addition to rain falling during the growing season. A wheat crop would require 275 to 325 mm (11 to 13 inches) of water, which would typically be a combination of approximately 175 mm available in the ground at planting and 200 mm rainfall during growing (with surplus being lost through evaporation in the early weeks of the season).[29] Sufficient moisture would normally have been available in the eastern Prairie Provinces, but even eastern Saskatchewan and Manitoba, for example, ran a 25 per cent risk of the amount being 50 to 75 mm less than average.[30] Such a deficiency would be a catastrophe for central and southern Saskatchewan and southeastern Alberta, but it would not have been a surprise on the dry margins of wheat-growing in the early decades of the twentieth century: much of that area would regularly have had a deficiency of 35 mm (nearly 1 1/2 inches) in a year.[31] Surrounding this dry heartland toward the north and west was a transition to more reliable precipitation. For example, a zone stretching from the Empress district, Alberta, in a curve to the southwest half way between Medicine Hat and Calgary, had a mean annual precipitation of 305 mm (12 inches), with 229 mm (9 inches) during the growing season.[32] While precipitation was seldom as critical on the boreal margin as in the southern parts of the western interior, Carder points out that for agriculture in the upper Peace River region "the greatest limitation is inadequate moisture in some years. Rainfall through the growing season is usually erratic," although there has never been a total crop loss because of drought in the region since records began.[33] The average annual precipitation at Beaverlodge, from 1915 to 1965, was 457 mm (18 inches), with 254 mm (10 inches) falling in the growing season; at Fairview the average was 454 mm.[34] Generally, on the north side of the Peace River basin precipitation is about an inch (25 mm) less than at Beaverlodge, but this difference is largely offset by a lower temperature and less evaporation. An ironic problem that the Peace River country shares with all the northern lands with severely cold winter temperatures is that much of the potential moisture provided by snow cover can be lost because at melting the ground is still frozen, allowing for excessive runoff.

Availability of sufficient water was the major factor determining the yield (bushels per acre) of the wheat crop, and in wheat country worldwide the hope is for at least ten inches of rain during the growing season. The impact on

Table 3.1
Wheat yields – average number of bushels per acre – for Saskatchewan (A) and the three Prairie Provinces (B), 1916–37

| *Year* | *A* | *B* |
|---|---|---|
| 1916 | 16.3 | 16.9 |
| 1917 | 14.3 | 15.6 |
| 1918 | 10.0 | 10.2 |
| 1919 | 8.5 | 9.3 |
| 1920 | 11.3 | 13.9 |
| 1921 | 13.8 | 12.6 |
| 1922 | 20.3 | 17.7 |
| 1923 | 21.3 | 21.7 |
| 1924 | 10.2 | 11.2 |
| 1925 | 18.8 | 18.6 |
| 1926 | 16.2 | 17.5 |
| 1927 | 19.5 | 21.2 |
| 1928 | 23.3 | 23.5 |
| 1929 | 11.1 | 11.6 |
| 1930 | 14.0 | 16.0 |
| 1931 | 8.8 | 11.8 |
| 1932 | 13.6 | 16.0 |
| 1933 | 8.7 | 10.4 |
| 1934 | 8.6 | 11.3 |
| 1935 | 10.8 | 11.3 |
| 1936 | 8.0 | 8.6 |
| 1937 | 2.7 | 6.7 |

Source: Stapleford, *Rural Relief*, 24.

wheat yield of the variability of precipitation, and especially of the increasingly recurrent drought conditions after 1928, is starkly illustrated in table 3.1.

This table reflects returns from the whole farming area of the provinces, including both the difficult and the well-endowed sections. More to the point of this review was the experience of thirteen municipalities in the persistently dry area of southcentral Saskatchewan, where the average yield of wheat per acre on a large sample of farms for the whole period 1921 to 1936 varied from 10.8 to 13.7 bushels; but, between 1929 and 1936 these municipalities averaged 4.3 to 8.0 bushels per acre, overall only 28.6 per cent of the production of 1921–28 when the averages were between 16.6 and 21.5. 1931 was a particularly bad year, when 85.8 per cent of the 1,117 sample farms harvested fewer than 6 bushels per acre of wheat; and 1937, when much of southern and southwestern Saskatchewan harvested no wheat at all, was disastrous.[35] Growing wheat – the normal ambition of the inter-war pioneer farmer – was a highly hazardous employment in the semi-arid areas even though (or perhaps *because*) moist climatic aberrations could bring heavy crops from time to time. During the difficult years comparisons would have been made with the boreal townships, where regular wheat harvests of up to thirty bushels per acre were being reported.

Just as the characteristics of the climate in the marginal areas showed critical variability from time to time, so the surface features and the qualities of the soil varied significantly from place to place. Omnibus terms such as boreal margin, dry belt, clay belt, or Prairie Provinces suggest homogeneity, but each was a conglomerate of surface conditions and soil qualities. The boreal and the dry margins showed marked differences: the former being characterized by tree growth (often heavy) and more permanent surface water; the latter being grassland, with large stretches of debris left from the dying of glaciers, landforms modified by persistent winds, and a deficiency of permanent surface water except for the exotic South Saskatchewan River and a few sizable lakes. There is an irony, however, in the seasonal variations in the water budget on the prairie (most obvious when viewed from the sky). In the spring thaw, with the winter snow cover melting on the still frozen ground, the prairie presents a watery scene, with a multitude of shallow pools scattered liberally across the flat land (occupied for a short, riotous interlude by crowds of migrating water birds); but by July the water has gone and the panorama has returned to its normal water-deficient condition.

The fertility of the soils on the boreal margin across the country was a virtual unknown at the beginning of the twentieth century, and had to be worked out largely by trial and error (the striking complexity of the soils and surface conditions is illustrated by Canada Land Inventory maps in chapter 5).[36] On the colder fringes the land could vary from the forest podzol soil, with an A (surface) "horizon" or "epipedon" containing decaying vegetation, to bands of gravel and sand, to wet patches, all within the bounds of one quarter section (160 acres). What nutrients there were depended on the character of the vegetation, and over time the leaching process moved the nutrients down through the soil layers to where tree roots, but not most farm crops, could easily reach. A generally low fertility of the northern podzols had long been suspected, and tests had given support to this view as early as 1906. Soil testing by the Ontario Agricultural College, which served as background for Fernow's assessment of the settlement potential, had raised serious questions about the possibility of agriculture in the Great Clay Belt (later the Cochrane District) of northern Ontario: "The chemical analyses of 18 of even the 'more promising' soils shows several of them as 'undesirable, and none of them except No. 8 come up to the standard of a virgin soil.'"[37] This view was reinforced when the modern soil survey was being carried out in the area in the 1940s. Soil surveyor G.A. Hills, in a general assessment of the Great Clay Belt, said "only a small percentage … has an agricultural potential sufficiently high to compete with those lands already developed in Canada."[38]

The boreal margin was considerably varied, and its soils did not have a bad reputation everywhere. In one study, on the northern edges of farming in northeast Alberta where boreal forest abutted parkland, agricultural scientists were surprised to find that the fertility of the grey woodland soils seemed to be almost equal to the dark soils of the nearby parkland belt. The explanation was that at first "breaking" (i.e., initial ploughing) of any virgin soil there were accumulated nutrients. These would continue to benefit the crops, even on the grey wooded soils, "to permit a mixed type of farming to be carried on with results comparable to those obtained on the black soil. No doubt, in the long run, the farms on the black soil will enjoy a comparative advantage."[39] But the grey wooded soils in northern Ontario were not able to demonstrate even the discounted qualities of

Illustration 3.3
A not uncommon problem on the northern fringes (e.g., the "quagmires" of Porcupine Plain): unloading a mired wagon to save a horse and cargo.
*Source:* Canada, Department of Interior photograph collection, LAC, Patent and Copyright Office collection PA41362 (dated 1930).

those in northeastern Alberta: the "cold" (i.e., slow to warm in spring) soils of the clay belt were not ameliorated by the large amounts of poorly drained and water-covered land in the environs – "The one feature which impresses one most is the swampy condition of the country" – let alone by the challenging climate.[40] A more complex assessment in the 1950s of one of the best parts of the Great Clay Belt revealed that up to 50 per cent of the land was poorly drained, and probably not one in five of the farms was economically viable (see illustration 3.3). In addition, the area around the town of Cochrane was still more than half tree-covered in the late fifties, and even the line of the railway was as likely as not to be bordered by woodland.[41]

The spread northward was helped by the development of faster-maturing strains of wheat to reduce the frost threat, although this did not lead to farm stability in northern Ontario. The half dozen areas of farmland across northern Ontario reached their highest amount of crop land by the beginning of World War II, after twenty to thirty years of coming to terms with the natural and economic contexts. From that point the farming diminished. In Troughton's view of the peripheries across Canada, and especially those in northern Ontario, "Human response in the marginal zone has been based on less than adequate knowledge ... and on a lack of policy and planning."[42] Expansion continued in Abitibi, through the opening of a series of carefully planned parishes, mainly north of the transcontinental railway line. Québec was more resolutely attempting to extend farm settlement into the various arms of the clay belt, and achieved notable success for a time based on planning, increasing understanding of the natural conditions and appropriate crops, and copious government assistance.[43]

The alternative to the northern forest for the late landseekers was remnant dry grassland. It was generally more accessible and more easily broken as crop land, but it had been extensively taken up a few years earlier. The land on the dry fringes could vary from the naturally fertile dark brown grassland soil, to light and sandy shortgrass soil, with blowouts where the sward on light soil was too thin, to glinting white patches betraying the salts raised by evaporation from ancient marine strata. Soil surveys done in Alberta during the 1930s showed gradations from the poor, wind-eroded soils of the southeast corner, into gradually more fertile soils toward the west and especially northwest of the Cypress Hills, where dark brown soils could produce bumper crops when precipitation was sufficient. The poorest soils by and large were found where precipitation was most unreliable. In the southeast corner only 1,824 quarter sections (19.8 per cent) out of 9,216 revealed any cultivation at all in 1940. Most of the 1,824 had between 10 and 140 acres cultivated; 226 were completely cultivated; but 441 others had been abandoned.[44] For the neighbouring province, the *Atlas of Saskatchewan* shows most of the land south of Prince Albert to be among the best suited to agriculture in Canada, although large areas in the southern part of the province have a risk of rainfall deficiency or soils with poor water-holding capability.[45] When these undesirable characteristics were combined with light sandy soils, they led to the infamous conditions of the Dust Bowl with, as an observer reported, the land "approaching the sand dune stage. Wind erosion in this area has covered fences completely … and has changed the surface of the ground into deep pits and hummocks of sand. Quarter-sections may be seen in this area that have not been touched in four years on which no growth is yet to be observed despite the favorable moisture conditions of the current year."[46] Dryland farmers could not look to an improvement toward a lower water requirement for their wheat to compare to the frost-beating, famous Marquis wheat in the north. The marginality of the dry frontier made its presence felt on a different timetable from frost damage on the cold frontier, and the dryland conditions of the 1930s made the northern margin in the woodland, where at least precipitation was more regular, appear desirable to people who wanted to farm. A few thousand "dried-out" farmers migrated north for a refuge (see illustrations 3.4, 3.5).[47] As one of the re-locating wives recalled: "I looked out at the fields and pasture where I could see one of our horses pawing the sand trying to get a mouthful of grass … Russian thistles were piled ten feet high in fence corners, and held there with layers of sand which had blown … we talked of moving to a place where there was green grass and shade. It would be a veritable heaven after so much dust and wind."[48]

## HARDSHIP LAND

Notwithstanding the fleeting popularity of the northern fringe for an escape from aridity, both dry and boreal areas were marginal, and during the 1920s and 1930s many farmers who had ventured into these questionable fringes were forced to give up. Official records note that as early as 1926 over four million acres were vacant and abandoned in the three Prairie Provinces. In Alberta it amounted to 8.2 per cent of the farm acreage; in Manitoba 4.6 per cent; and in

Illustration 3.4
A graphic illustration of the conditions that drove settlers from southwestern Saskatchewan to seek moisture in the forest to the north.
*Source:* Saskatchewan Archives Board, photograph no. R-A4822.

Illustration 3.5
Moving from the vicinity of Morse, near Swift Current, Saskatchewan, heading north.
*Source:* Saskatchewan Archives Board, photograph R-A4287.

Saskatchewan 2.2 per cent.[49] The storied homestead of 160 acres (the quarter section) was proving too small, especially in marginal conditions. In the Prairie Provinces, the farm of a quarter section or less accounted for 74 per cent of the 19,000 unoccupied farms in 1926. Fourteen per cent of quarter-section farms were abandoned in the three provinces as a whole; in Alberta the figure was 19 per cent. The abandonments were noticeably concentrated in the southeast corner of Alberta and the Interlake and adjacent districts of Manitoba, representing respectively the dry and the boreal margins.

In addition to limiting agricultural possibilities, nature conspired to put hazards in the way of settlement. The shifting of settlement into wooded land, after a couple of generations in grassland, caused incidents of disorientation and difficulty of cross-country travel, similar to those on the eastern forest frontier. Whereas on the open plains the winds were legendary and combined with snow became life-threatening blizzards, in the wooded areas the trees interrupted the wind and as a result blizzards were rare, "not more than one being expected in 20 years" (upper Peace River region).[50] A wind and grassfire combination could also be extremely damaging, even fatal, given the astonishing speed it could reach: it was said that a prairie grassfire could outrun a rider on horseback. The western end of the boreal margin was periodically visited by a chinook, in which snow would be melted and plants could be damaged by rapid temperature changes. But a more characteristic experience of the northern fringes was exemplified by the upper Peace River, which recorded fourteen episodes in fifty years when the winter temperature remained below 0°F (–18°C) for more than ten days.[51]

It was no accident that land was still unclaimed after World War I, despite the energy of the nineteenth-century continent-wide agricultural expansion. These "unused lands" (in Murchie and Grant's terminology) were either too dry and recognized as more suitable for ranching, or too cold for most valuable farm crops, or too remote for participation in commercial marketing, or too wet or infertile for crops. They were by most measures "marginal," but held attraction for the land-hungry because of examples of rare farm success publicized by government or land company promotion. The people who went to the late frontiers were in many cases desperate – desperate to claim some of the last land thought suitable for farming, or desperate for *any* kind of livelihood rather than destitution or public charity. The various influences at work and the depth of emotion underlying this search for farms are described in the following chapters.

# 4

# Frontier Myth and Self-serving Agendas

Human purposes ... never take full account of the circumstances upon which they impinge, and every so often act as triggers, provoking results that were not imagined ... the more skillful human beings become at making over natural balances to suit themselves, the greater the potential for catastrophe.[1]

The frontier has been seen as a superorganic entity, much bigger than "the hither edge of free land," in all parts of the New World. Frederick Jackson Turner and his followers endowed the frontier with mythical powers, bearing on social interactions, individual freedom, positive thinking, the form and practice of "democracy," and so on. Turner's frontier has been severely tested and cut back to size by critics who found that the frontier experiences of many people and the pioneering conditions in many areas of agricultural expansion were not satisfactorily represented by its exuberant optimism. But government officials and nationalistic promoters have not been nearly as analytical as the critics, knowing that the slogans of the frontier and the expansion of settlement have had powerful reverberations in the popular mind, especially in the last two and a half centuries. Although government policies and legislation were the main expediters of the movement toward the fringes, they were not the only forces at work. People were yearning for land of their own, churches were seeking to expand areas of influence, divergent ethnic and religious groups were looking for refuges, railways were striving to increase the number of customers, and bureaucrats were co-opted by the idea of the frontier and its promise of progress. It was important to many people, for a variety of reasons, to *keep the frontier rolling on.*[2]

## MOTIVATIONS

### *Settlers*

Late-arriving land seekers were as wedded to the frontier dream as their predecessors had been. It had always been possible in living memory, and for a considerable time longer than that, for a person to head off to new land in the New World. Fathers, uncles, or acquaintances had done it, and the opportunity was taken for granted. And it appears that outside the New World countries, and beyond the range of first-hand experiences and up-to-date information about problems as well as opportunities, the frontier dream was even less compromised: the further away the dreamer, the rosier the dream – at least in many cases. People without access to the main body of information in English were at a special disadvantage. Quite often their chief source of information was someone who stood to gain by persuading them to emigrate, such as an agent paid by a government or a shipping company. One example was Joseph Oleskow, who worked as the Canadian Immigration Representative for the Austro-Hungarian Empire in the 1890s, principally in what are now the Ukraine-Poland borderlands. He declared in his booklet *On Emigration* that "Each worker in America, even a labourer engaged in the dirtiest kind of work, is, by our modest standards, a gentleman ... I would like to emphasize particularly the cleanliness of Englishmen and Americans: they would never sit down to eat without washing before the meal and they all wash with soap at least three times a day." The accuracy of his views is brought into question by his claim that "There is scarcely any hard tim[b]er woods in Canada, excepting dwarf oak trees in Manitoba and the so-called tamarack ... It is best to settle along the rivers, streams and large lakes."[3]

The attraction at a distance was reflected in the composition of the population (as highlighted in chapter II) especially in the western provinces where a large portion of the population in the marginal areas was not born in Canada. In contrast, more mature pioneer districts such as Algoma and Timiskaming in Ontario, and all the rural parts of Québec, in 1921 were even below the national average of 29 per cent of the residents not born in Canada (this includes "British" and "foreign-born"). The Rainy River district in northwestern Ontario, with 34 per cent of its population born outside the country, signalled the different demography of the West. The Prairie Provinces in general had been relatively more attractive to farm immigrants, and the further west the more so, as indicated by the steady increase of the immigrant proportion of the population of the fringes from Manitoba to Alberta. For example, Minitonas, in Census Division 15 of Manitoba, had 33 per cent of its population born outside Canada, and the northerly Divisions 14, 15 and 17 in Saskatchewan had respectively 40, 37, and 43 per cent born outside (see figures 1.1, 1.2, and 2.1 for locations). These latter figures pale beside those for Alberta where Division 9 had 56 per cent, Division 14 had 46 per cent, and Division 16 had 47 per cent non-Canadian-born residents.[4]

In the decade leading up to 1931, the proportion of the population of Canada born outside the country had reduced to 22 per cent. The pioneer areas of Québec continued to be almost entirely born in Canada, the figures for Abitibi and Lac St Jean being between 96 and 99 per cent. In Ontario, the Cochrane District, formed from northern parts of Timiskaming and Algoma, had 28 per cent of its people born outside Canada, although individual township percentages ranged from single-digit to slightly above the district average. The West remained a haven for the immigrant. The figure for the non-Canadian-born in Minitonas had jumped to 44 per cent, and neighbouring Swan River was 30 per cent. The population reports of the northern areas of

Saskatchewan showed 34 per cent of Division 14, 33 per cent of Division 15, and 38 per cent of Division 17 born outside Canada, although some of the townships had over 40 per cent. Alberta's Division 9 once again showed the highest rate, with 47 per cent born outside the country, while Division 14 had 43 per cent and Division 16 had 42 per cent.[5]

It was common for writers to talk about "land hunger," and certainly from the beginning of the twentieth century to the 1930s there was a continuing demand for land to farm. Along with the pressure from immigrants, notably a portion of the million or so who came in the decade before World War I, followed by a sizeable contingent of British soldier settlers after the war, the main demand came from inside the country. One desperate search for new land came within the Prairie Provinces, where there was a well-documented flow from dried-out margins toward the northern fringes.[6] Even before the war, the harvest excursions from eastern Canada to the Prairie Provinces, as well as copious discussion and depiction in the popular press, provided enticements to people hoping to have a farm of their own.[7] In Québec and Ontario, and on a minor scale in New Brunswick and Nova Scotia,[8] the search would often stay relatively close to home – higher up slopes than previous farms or in isolated pockets of eligible soils. But for groups like the Bruce Preparedness League, for which Saskatchewan's Deputy Minister of Agriculture could hold out little hope for good homestead land as early as 1918 (see chapter 1), and for thousands of individuals, the search was much farther afield. It was important for these late frontiersmen that the idea of frontier remain alive. It was probably important too for the society at large, in the sense that cheap land had been a perennial component of the concept of the New World for at least a century and a half. Would the passing of the frontier mean the death of optimism, the onslaught of economic depression? Frederick Jackson Turner and some historians before him thought so, and politicians in various parts of the New World clearly did as well, because they continued to make it possible for crop farmers to expand even into areas shown by science to be unpromising. Some politicians sparred verbally with scientists they considered to be "obstructionist" for predicting limits to the expansion.[9] The politicians were responding in part to the keen desire of many of their constituents for the frontier to be kept open. For one thing, this would allow aspiring farmers in the older-settled areas to retain the hope that, although their home district was "filled up," the traditional opportunity to find new land would still be available.

The keen desire for land was well documented in the testimonies given to the Saskatchewan Royal Commission on Immigration and Settlement. At the hearing in Prince Albert in April 1930, Julius Androchowiez recounted his activities on behalf of the Canadian National Railway and International Mercantile Marine Co. In the previous three years he had located dozens of families, primarily eastern Europeans, north of Prince Albert on land that was awaiting clearing or had been abandoned. He gives the example of a Pole who came with very little, but "worked for two months hunting up homesteads" and after four years of hard work had achieved a successful farm.[10] Farther northeast of Prince Albert, the secretary of the United Farmers of Canada at White Fox illustrated the demand for land in that district: "The number of homesteaders there last year was between five and seven hundred ... They went into Townships fifty-two and fifty-three, Range eighteen, Townships fifty-two and fifty-three, range nineteen and Township fifty-one, Range nineteen. That is where the principal settlement has been made. These were townships the Government survey showed *were not suitable for farming* ... I was through the Russian settlement at Meath Park and Foxford last year ... where they settle in colonies they don't intermingle ... They don't know anything else and don't learn other customs ... When a settler goes on this land which is unsuitable, the agent tells him it is unsuitable, but he goes just the same."[11] Apparently these immigrant settlers would prefer to listen to Joseph Oleskow than to the local expert.

The keen desire is probably expressed best by novelists who observed the driven expectations of the new homesteader.

Frederick Philip Grove was one writer who lived among the land seekers and captured the search in his novels, with names such as *Fruits of the Earth*, *Settlers of the Marsh*, and *Over Prairie Trails*. His epic settler was Abe Spalding who "was obsessed by 'land hunger'; and he dreamt of a time when he would buy up the abandoned farms ... he would conquer this wilderness; he would change it; he would set his own seal upon it! For the moment, one hundred and sixty acres were going to be his."[12]

*Governments*

The most active provincial government in planning settlement was Québec. By the second half of the nineteenth century the thoroughly planned expansion of farming areas within the province was becoming a main pillar of the effort to save French Canadian culture and presence. The tradition of setting up Québécois agricultural colonies on newly discovered pockets of tillable land inside the province had been nurtured by the Roman Catholic church, and in the second half of the nineteenth century there was a great deal of experience to draw on. Because of the large number of French Canadians leaving the province for employment, political motivation for expanding the colonization initiatives was also strong. In the twentieth century the lead was usually taken by governments, with thorough collaboration between the provincial government and the church in Québec. The European geographer, Pierre Biays, catches the spirit of the back-to-the-land movement in Québec after 1910 in describing this "official colonization," primarily government directed, as "elevated to the status of a cult, of a national institution."[13] The Québec government had introduced a Department of Colonization in 1921 and in 1923 instituted a system of bonuses for work done by new settlers to develop their farms.[14] A colonialist mentality was not restricted to Québec, as exemplified (below) in Clifford Sifton's attitude to the European "peasants" he spread on marginal lands. Many of the administrators overseeing boreal margin settlement seem to have been engaged in a kind of internal colonialism.

The opening of northern settlement peripheries in the twentieth century, up to the time of World War II, was essentially the last spasm of expansion of the agricultural frontier even in Québec. Interest in finding new areas was ongoing, with the church's concern over the secularizing power of the city providing a regular stimulus to popularizing the rustic ideal. The pressures of the depression after 1929 also quickened the federal government's interest in placing economic casualties on unoccupied land with some potential for crop growing. Thus the federal government devised a scheme that came into effect in 1932, in which the federal, provincial, and municipal governments collaborated in putting people on farm lots (following primarily emergency relief acts in 1930, 1931, and 1932). The federal government initiative, known as the Gordon plan (properly entitled "An Act respecting Relief Measures," which, like its contemporary "An Act respecting Unemployment and Farm Relief," was one of a series of relief acts in the 1930s leading to the unemployment insurance act of 1940),[15] provided the Québec government an opportunity for launching into a renewed effort at colonization inspired, in Serge Courville's view, by widely publicized regional planning in the United States.[16] The Gordon plan, administered by the provincial Commission "du retour à la terre" (back-to-the-land), sent out about one thousand families, mainly to Témiscamingue and Abitibi, of which a quarter returned to the towns after only a few months.

In 1935 the Québec government turned from the Gordon plan to a plan of its own in which the clergy would be involved, replacing the collaboration with the federal government. The provincial initiative, under the name Vautrin, claimed itself to be a "national enterprise" engaged in providing a solution to unemployment and correcting the numerical imbalance between urban and rural populations.[17] This became by far the largest settlement on new lands at this time in Canada, locating over 55,000 persons in the three years to 1937. There were some clear preferences for the candidates: first choice were sons of farmers (7,853 household heads); second were those prepared to participate in integrated colonies such as had been developed to a high level in Québec (7,419 household heads); and in the "others" category were nearly 3,000 (including 1,500 farm labourers). Even with this elaborate and well-financed plan, Courville reports that more than one quarter of those placed on the land abandoned their locations.[18]

The Vautrin plan was overlapped, in 1936, by a new federal act, the "Unemployment Relief and Assistance Act" (*Statutes of Canada* 1936, chapters 15, 46), which invited provincial collaboration. This gave rise to a fresh arrangement between the Québec and federal governments, called the Rogers-Auger plan, to be managed by the provincial back-to-the-land commission. This settlement scheme was directed only to unemployed men who had at least a rudimentary knowledge of farming and a wife who could assist with farm work. Various kinds of financial assistance were provided, but the abandonment rate ranged between 30 and 40 per cent. In a review of thirty years of settlement overseen by the government in Québec, from 1910 to 1940, Gosselin and Boucher calculate that 51.4 per cent of the 80,175 farm lots sold "were revoked or cancelled for non-performance of settlement duties," and for every settler that stayed on the farm approximately $2,000 of government money had been invested.[19]

Québec was the *ne plus ultra* of assistance to marginal settlement in early twentieth-century Canada and, although the proportion of abandonments there was substantial, in comparison with almost all other parts of Canada the Québec move north was relatively successful (see a comparison with Ontario in figure 4.1). Other provinces, however, did involve themselves from time to time in helping needy citizens to locate on unused or abandoned land, while incidently extending the provincial ecumene. A.R.M. Lower reports a government official's opinion that re-echoed in the twentieth century and in jurisdictions in many parts of the New World:

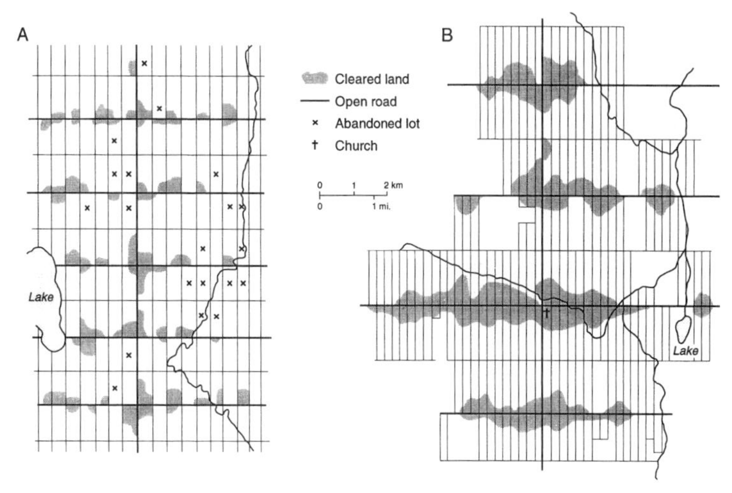


Figure 4.1
Simulation of contrasting landscapes resulting from twenty years of provincial settlement policies: parts of A) Glackmeyer Township, Cochrane District, Ontario, and B) St Henri de La Morandière parish, Abitibi District, Québec. Québec had developed a colonization system that usually led to a more integrated settlement, whereas the Ontario approach was primarily to fund the building of roads and bridges. Information on abandonments is not available for the St Henri area.
*Source:* Based on Biays, *Les Marges*, fig. 30 and fig. 40.

"the remaining public lands are admittedly not good but ... should be disposed of as quickly as possible, the settler being 'of infinitely more importance to the country than the land.'"[20] The mystique of the frontier remained strong in the twentieth century, and a desire to further expand the settled area motivated often ill-advised political decisions.

In the 1930s New Brunswick gave some assistance to persuade settlers to move onto wooded land that was generally at a higher altitude than previously farmed land.[21] This New Brunswick frontier provided a classic example of what was a typical, complex relationship between the agricultural and the forestry activities on the cold margins across Canada. The surrounding forest could offer an immediate saleable product or employment with a lumber company, or credit on winter supplies on promise of delivering pulpwood to a dealer, sometimes to the serious detriment of farm improvement. At other times, when the timber resource was getting scarce and when settlers were clearing land by burning, the farmer and the forester could be "at daggers drawn."[22] (See further discussion of agro-forestry in chapter 5).

Some provinces, while generally reluctant to break custom and dole out money to entice farm settlers into new areas, came on board with the federal government when it introduced the Gordon plan. The Ontario government joined the triumvirate – federal, provincial, municipal – by contributing its third to financing the Relief Act of 1932, although it seemed a rather hesitant participant. Biays characterizes the settlement in northern Ontario after the opening of the twentieth century as pursued in an individualistic and Anglo-Saxon ("liberal") spirit, unlike Québec's colonization imbued with cultural survival: in Ontario "there has never been a question … of social philosophy."[23] No social philosophy in the strict sense may have been true enough, but certainly a political philosophy was at work in which laisser-faire was the recurrent doctrine (see figure 4.1).

Peter Sinclair cautions against seeing the approaches to the northern settlement in Ontario and Québec as entrenched opposites. He argues that the intentions of the two governments were not always starkly different, especially in light of the periodic changes in conditions and the parties in power.[24] The pressure of welfare costs in the 1930s did persuade the Ontario provincial government of the need for some relocation assistance, but unlike large expenditures for building roads and bridges in the north, grants to individual settlers were seen as an aberration. The fractious debate over moving welfare recipients from cities to rural land was copiously fueled by some schemes related to the Gordon plan. The most notorious was the shipping of twenty-eight welfare families from the city of Windsor, Ontario, to the Great Clay Belt in 1932. Apparently the preparation and coordination were seriously deficient, especially where the character of the land and climate were concerned. Many of these people were very quickly in a state of near starvation because of the insufficiency and tardiness of supplies, the lateness of the spring, and the lack of farm equipment. A sympathetic cleric in Kapuskasing said:

> It's absurd for the government to bring these people here and expect them to become self-supporting in two years on $600, especially on the land on which they were placed. Even the old settlers with scarcely an exception are on direct relief. These people have no chance: without tools or equipment what can they hope to accomplish? With assistance they might clear and plant enough to produce enough for next winter but with the lack of interest shown on the part of those responsible I don't quite know yet what the winter will bring to those poor people.[25]

Especially because of a circuit through the Clay Belt by the mayor of Windsor and the negative publicity it generated, the plight of these settlers received a great deal of attention and the provincial government provided some additional assistance.[26]

Ontario did not have a well-coordinated system for expanding settlement into new areas. Although the Department of Lands and Forests had the major responsibility, practical activities were divided among various branches, and the hands-on dealing with farm settlers fell to the Department of Agriculture's Colonization and Immigration Branch. Some "free grant townships" still remained in Ontario, but the Clay Belt lands in Cochrane District, as in Timiskaming, were to be sold. Unlike the survey across most of the north, where farm lots were 160 acres, in Cochrane and Timiskaming the standard lot was to be 80 acres, as an encouragement to closer settlement. Additional land was available to a settler, but familiar settlement duties – of residing on the lot, building a house, a modest rate of clearing, and citizenship – were required on the 80-acre farm before a deed would be issued on any property claimed by the settler. In an omnibus overview of settlement on new land in Ontario from 1912 to 1937, Gosselin and Boucher calculate that nearly two-thirds of the lot entries were cancelled on account of failure to complete settlement duties, even in townships where the land was sold.[27]

The Ontario approach to northern settlement, for which the Northern Development Board was established in 1912, was motivated by business principles more than by a desire to colonize, thus explaining the heavy weighting of government expenditure in the north on roads and bridges that could also encourage forestry and mining. Very little financial assistance other than loans was provided to settlers, except under the Gordon plan, when the province shared the cost equally with the federal government and the municipality from which the relief settlers came. The differences in the settlement resulting from the policies of the governments of Ontario and Québec are encapsulated in figure 4.1. The Ontario picture is of scattered farmsteads, isolated in small pockets of cleared woodland, reminiscent of what the scene might have looked like at the beginning of settlement in "Old" Ontario early in the nineteenth century. The sad truth, however, is that in New Ontario most of the clearings did not expand to the dimensions of a proper farm. The Québec scene, though undeniably rustic, depicts conditions promising much more social cohesion and land use interconnectedness – the ingredients that encouraged more permanence, certainly through the first generation.

Unlike the eastern provinces, which were attempting to open new areas after a hiatus in which the agricultural frontier had moved elsewhere, the Prairie Provinces were still riding on a wave of immigration at the opening of the second decade of the twentieth century. People were still confident that good land was continuing to be found. The three provinces, two of them very new, relied on that confidence and on the expectation of it being maintained into the foreseeable future. Urban areas were also growing at a great pace, but although the urban real-estate bubble burst in 1913, the search for farmland continued even in reaches far from transportation and markets.

The federal government, especially under the Laurier Liberals, actively promoted the settlement of the West and for that reason retained administration of Crown lands until

1930. In holding off western demands for the Crown lands at the time of the establishment of Saskatchewan and Alberta in 1905, the federal government inflamed a budding "regional grievance," in Owram's phrase, that has continued to bedevil intergovernmental relations.[28] As Minister of the Interior, Clifford Sifton championed initiatives to bring settlers from overseas, his opinion being that:

> In northern Ontario, Manitoba and Saskatchewan we have enormous quantities of land perfectly fit for settlement. These are not lands on which the ordinary Englishman or American will go, but ... I scattered a number of European peasants on these lands, and those are the only parts where the people are not in debt ... you have to put these men who will be satisfied with the standard of living associated with that class of country there[,] or leave the land untilled.[29]

Despite the well-founded uncertainty of the agricultural usefulness of much of the land, the government had no intention of putting up barriers to settlement or even of scientifically monitoring the process. In 1907, in Alberta, the surveyors were being sent to survey land for settlement in areas north of the previous boundary of accepted agricultural incursion. Planning for a railway soon followed and, after only seventeen years in which entry was overland via the tortuous Edson Trail or the equally difficult trail from Edmonton along the south side of Lesser Slave Lake, settlers were able to enter with their equipment by train even west of Grande Prairie in the Alberta Peace River country. The federal government began establishing agricultural research stations in the West, where different crops and rotations were tested, eventually including Beaverlodge under W.D. Albright.

The provincial governments in the western interior were anxious to attract population, but the management of Crown lands remained in the hands of the federal government until 1930, so the provincial role was mainly to carry out research and offer advice through its Department of Agriculture. Prior to the various federal-provincial initiatives of the 1930s, the most substantial move of the federal government was in its soldier settlement schemes beginning near the end of World War I. The first version was laid out in the Soldier Settlement Act of 1917 after extended discussion in which, according to Fedorowich, some of the provinces had to push the federal government to pass legislation.[30] Apart from the popular pressure to reward servicemen for their sacrifices, a persuasive argument with many politicians was that the scheme would provide population for rural areas, many of which had lost out to the strengthening pull of urban places. Soldier settlers from the Canadian and other British Empire forces were eligible to apply for land and loans on favourable terms and, where Dominion Crown land existed, they could apply for a "free" homestead – the famous $10 bet with the government. It turned out that a much larger proportion of the available land than expected was either unusable or more than fifteen miles from a railway – "85 percent of the twenty-two million acres of vacant dominion land in the west" – and as a result between 1917 and 1919 only a little more than 2,000 soldiers took up the opportunity.[32]

Murchie and Grant claim that, in comparison with earlier permissive legislation surrounding the spread of settlement across the country, the act of 1917 was the "first real attempt at a land settlement policy." The emphasis should be on the word "attempt," because it was found necessary to pass another Soldier Settlement Act in 1919 that adapted the policy more closely to what returned servicemen needed, namely money to borrow on easy terms for a greater variety of farming applications.[32] This second version was more successful: in Manitoba alone over 3,700 soldiers took advantage of its terms to go onto the land (although after eight years 43 per cent, for a variety of reasons, were no longer on the chosen location).[33] Most of the properties offered for settlement were probably what Murchie and Grant referred to euphemistically as "unused lands," including areas previously abandoned or deemed unsuitable, as well as some not covered by earlier surveys (figure 4.2). A major extension of new land was found in the northwest corner of the Manitoba

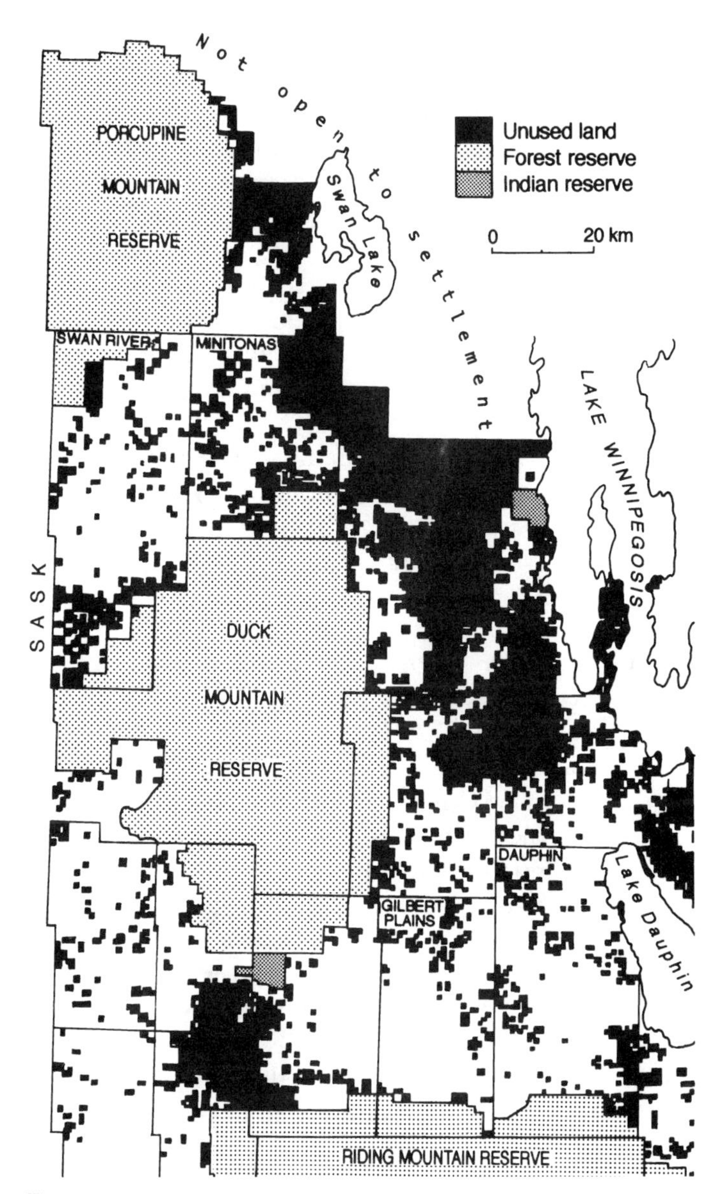


Figure 4.2
"Unused" land, western Manitoba, 1926.
*Source:* Redrawn from map in Murchie and Grant, *Unused Lands of Manitoba.*

ecumene, from the northern townships of Minitonas and beyond into unorganized territory along a railway recently built around Porcupine Mountain and into Saskatchewan. Saskatchewan and Alberta each had nearly twice as many soldier settlers as Manitoba, the three provinces together accounting for well over half the approximately 27,500 spread across Canada.[34] By 1923 soldier settlement and its bureaucracy began to be blended into the general management of immigration by the federal government. In the decade following World War I, government involvement in settlement in the West, including some in marginal areas, emanated primarily from Ottawa, with a lesser initiative from London in aid to war veterans and the British Family Settlement Scheme.

Saskatchewan organized its Commission on Better Farming at the end of World War I, and employed agricultural representatives who gave advice to farmers and wrote weekly reports to the Deputy Minister of Agriculture. One relatively early example from the dry belt around Mortlach, Saskatchewan, an area from which a large number of farmers headed for the northern margin a decade later, reported in October 1921, that:

> During the past week I have covered a considerable portion of the blown area around Caron and Mortlach … In taking a general tour … to determine the boundries [sic] of the drifted land I found the roads even now quite difficult for motor traffic. A few of the roads especially leading west where the farms are now vacated are almost impassable. There are several sections in a line that are unoccupied, and entirely over-grown with Russian Thistle. For the most part this takes in the higher ground.[35]

The agricultural representatives provided written snapshots of general conditions in the countryside. A report in 1919, from far into Palliser's Triangle, gave an early clue to what would become a primary solution for farmers "dried out": "I met several 'prairie schooners' on the Trail at Herbert containing families and personal effects, with stock driving behind. On the 14th inst. a number of the 'progressive'

Mennonites are gathering between Neville and Rush Lake to travel north."[36]

In the 1930s the Prairie Provinces participated in the same way as the eastern provinces in the so-called relief land settlement schemes. Now with the ability to open their newly acquired Crown land, and to survey for settlement, they could expand the area of available land beyond that of the soldier settlement and further into the northern margins. An indication of the increasing pace from the end of the 1920s is found in Robert England's almost on-the-spot report that "In the past seven or eight years 30,000 families have taken up farms in the more northerly areas of Manitoba, Saskatchewan and Alberta."[37] This was largely settlement from within Canada: responses to a questionnaire in Stapleford's inquiry into the operation of relief measures in the mid-1930s indicate that while a majority of those abandoning the drought areas were leaving the Prairie Provinces altogether, over a third were shifting to the northern fringes; and the federal and provincial governments were helping, as Stapleford reports for Alberta, where "an arrangement was made with the railways – the Canadian Pacific Railway, the Canadian National Railway and the Northern Alberta Railway – for a special tariff of approximately two-thirds of the normal rate for the removal of farmers on poor lands to more favourable areas."[38] The Swift Current *Sun* in 1934 illustrated the attraction of the north for local farmers hard hit by drought: two men had recently returned; "They were looking over the north and report lots of feed in that country" (recall illustrations 3.4 and 3.5).[39] Almost from the opening of the Peace River country just before World War I, a significant proportion of its settlers were getting away from the dry prairie further south in the western interior of Canada or the United States. It is important to remember that although the boreal margin had its problems, they represented only a small part of the crisis of the thirties. To a significant degree the north served as an antidote.

In addition to rural relief, there was a comparable need for urban relief in the western cities. As in the eastern provinces, it was believed that the plight of the urban destitute might be solved by placing them on a plot of land where they could at least grow their own food: as it was phrased, "helping people to help themselves ... would be in the public interest, both from the stand-point of the families assisted and the Canadian taxpayers who are called upon to shoulder the burden of relief costs."[40] There was a passionate attempt by a society still tied to a farm past, and with no assurance of industrial alternatives, to push back the juggernaut of urbanization. The Canadian population had proved to be more urban than rural by the census of 1931 and, in the despairing situation of the depression, it was easy to presume the crossing of that threshold to be somehow responsible. But urbanization was to continue its inexorable growth, and a large part of the explanation for abandonment of farms – especially by the rusticated urban relief recipients – came from the realization that if opportunities were going to materialize, it would more likely be in the cities than in marginal farming areas. In addition to the efforts to persuade the urban destitute to move to the country, relief continued to be provided in urban places from city to village size during the thirties.

The federal government had offered aid to the Prairie Provinces from time to time from the first decade of the twentieth century – significantly in 1908, 1914, 1919, and 1920. But the earlier rescues and the legislation that had dealt with conditions in the teens and twenties were dwarfed by the scale of the catastrophe of the thirties. By far the hardest hit province was Saskatchewan, the province most dominated by agriculture, and the scale of the aid given to it is astonishing. With the 1931 crop failure, the federal government pumped about $22 million into the Prairie Provinces in price support (more than half as a five cent per bushel bonus to wheat-growing farmers), relief, and agricultural aid. In the following five years the need for aid varied, at a lower level, until 1937 when the crisis reached the highest level because of an even worse crop failure, and in Saskatchewan the total number of people receiving relief in some form totalled 426,086 in a provincial population of 930,893.[41] The other two Prairie Provinces required less relief than Saskatchewan. For example, in the 1937–38 fiscal year, when Saskatchewan received $26 million in aid from the federal government, Alberta received $3,516,774 in various forms of relief, including work-as-welfare schemes, feed and fodder, and direct money grants. In 1935 the area allocated to the ministrations of the Prairie Farm Rehabilitation Act – the federal government legislation designed to meet the distress and damage caused by the extended drought – embraced 81,576 of Saskatchewan's 142,391 farms, 35,390 of Alberta's 100,358 farms, and 15,194 of Manitoba's 57,774 farms.[42]

During this difficult period when the federal government was providing major financial aid to western farmers, the provinces were also doing whatever they could. The Saskatchewan government, for one, was contributing as much as or more than the federal government, ranging from a low of $3.3 million in 1932–33, to nearly $19 million in 1931–32 and 1936–37, to a high of $21,297,510 in 1934–35. The Alberta government was also involved in a range of aid programs, though the sums of money were considerably lower. Rural relief in Alberta, including all but the largest cities, rose to a maximum in 1935, when nearly 14,000 families and 2,564 single persons were helped. The numbers diminished steadily in the following two years. The federal government took a major responsibility for relief in Alberta in the renewed drought conditions of 1937, including 100 per cent relief funding in fifty-two municipalities in a large block in the southeast portion of the province; it also provided a monthly grant to bolster the province's relief fund. In Manitoba, as in the other Prairie Provinces, the federal government covered 100 per cent of the relief in the A (or severe drought) area of the southwest, where a half dozen municipalities had suffered five straight crop failures by 1936. In the rest of the province relief was generally split three ways, between the two senior levels of government and the municipality. As the Stapleford report phrased it in 1939, "While the total costs of relief reach staggering figures, the principle is everywhere accepted to-day that provision must be made for those in need."[43] Note, however, that most of the agricultural aid was a loan secured by a promissory note, although cancellation of farmers' debts to government did occur.

A major role of the federal government during the thirties was to provide the money for all kinds of relief, because even provincial governments had difficulty borrowing money. Saskatchewan was a prime example, where 90 per cent of the $110,602,638 devoted to relief in the six years up to spring 1937 came from Ottawa, nearly $36 million of which was an outright contribution. Alberta incurred relief expenses between April 1930 and March 1938 of nearly $59 million, of which the federal government contributed just over $17 million.[44]

The western provincial governments had provided emergency relief long before the crisis of the 1930s. Basic legislation arose out of the crop failure of 1919 and resulted in proposals in Saskatchewan and Alberta to ameliorate and head off such situations in the future. In 1920 the Saskatchewan government introduced The Municipalities Relief Act, and in 1931 it set up an agency called The Saskatchewan Relief Commission. The early Alberta government response was to establish in 1921 The Survey Board for Southern Alberta, with a mandate "to inquire into, report on and make recommendations ... in regard to ... matters affecting the welfare of those areas ... which are subject from time to time to drought."[45] Continuing serious droughts in southeastern Alberta resulted in An Act Respecting the Tilley East Area in 1927, followed by similar legislation for adjacent areas.[46] The Alberta government, as other provincial governments, collaborated during the 1930s with the federal government under The Relief Act of 1932 and its later iterations.

Politicians and civil servants in federal and provincial governments clearly took for granted the continuing expansion of rural settlement into new areas, either as a natural extension of the nineteenth-century frontier or, more cynically, as a convenient valve for social pressures. Thus, almost as guinea pigs, settlers were allowed into areas that, according to available scientific evidence, promised a broad range of difficulties if not failure. And even when enlightened officials disseminated useful information and attempted to create safety nets, thousands of cases needed rescue from ill-fated locations. The natural environment for pioneering after the first decade of the twentieth century was profoundly different, one might say threadbare, compared to the rich tapestry of opportunities, especially in the western interior, one generation earlier.

*Religious Denominations*

Among the most cohesive and successful experiments in settling the margins were those fostered by religious organizations. Some small religious denominations were seeking refuges, and areas that were avoided by the main concentrations of rural settlement could appear suitable. Representatives for 12,000 conservative Manitoba Mennonites, in one little-known example, entered into negotiations with the government of Québec and carried out on-the-spot investigations in that province in 1920. The Mennonites had run afoul of the Manitoba government, especially over their desire to determine the schooling for their children. Having decided to leave Manitoba, their leaders considered the possibility of finding up to 400,000 acres in the new territory of the Abitibi clay belt. The news of the Mennonite investigation of the northern areas raised interest even in Montréal, where the major French-language newspaper pointed out that "We have in the Abitibi and Temiskaming millions of acres of still unoccupied land, waiting for those who would clear it ... These lands are in large measure covered by tree species which command enormous prices ... which makes them a precious means of subsistence for new settlers as they wait for the land to begin to produce crops."[47] Rather than an invitation to the Mennonites, however, this article was designed to rouse French Canadians to the settlement possibilities in the area. Although the Mennonites were entertained politely by Québec officials, this group decided to join other Mennonites from western Canada and settle in Mexico.

The religious groups that sought a refuge, that is the Mennonites, Hutterites, Doukhobors, Mormons, and Jews, were motivated primarily by non-economic factors. Much more than in initiatives sponsored by some established churches and the government, most of these minority religious groups were pushed out of their previous locations by increasing hostility, even persecution, and restrictions on their ways of life. The Mennonites had settled in various parts of the western interior of Canada toward the end of the nineteenth century. They were particularly well represented in southern Manitoba, in the East and West Reserves, in central Saskatchewan, and north of Saskatoon, where Robert England observed their expansion farther toward the northern fringes in the early 1930s.[48] Their success in farming led to expansion across the West, sometimes in block settlements, sometimes as independent farmers, depending on the degree of exclusiveness of their version of the beliefs.[49] The independents gradually became indistinguishable from the main body of settlers, but some block settlements continued and in most cases flourished. Most of the Mennonite influx occurred early enough that good land was still available, and most of their locations were in the centre or southern sections of the western interior, not in marginal conditions. But Mennonite groups looking for new settlement areas after 1920 had to consider more and more marginal locations, including the boreal forest. Some of the latest moves of farming members of this religious persuasion have been to an extreme northern outlier, La Crete, at more than 58° north, near Fort Vermilion on the Peace River.[50] The Mennonite tradition had been "to remain separate from the materialism and godlessness that they associated with the larger prairie community," by carrying on their rather closed village-based farming; but, by World War II "The village agricultural system and the separate school were things of the past; the language and the faith remained."[51] Mennonite farmers were moving out of the villages to live scattered on their own rural properties.

The Doukhobors – officially the Christian Community of Universal Brotherhood – were another group wedded to the land. They entered western Canada at the turn of the twentieth century, bringing a belief system that spurned material show, swearing of allegiance, and submitting of vital statistics, but embraced communal ownership of land and labour, pacifism, and vegetarianism. Although they did well on the land, and established dozens of agricultural villages in less than a decade in a rather forbidding section of northeastern Saskatchewan, their refusal to claim individual homesteads and to swear allegiance to the Crown was met with intransigence by the authorities. These difficulties in what was to be their promised land led to divisions in the group, so that before the second decade of the twentieth century many Independents had split off from the Community Doukhobors. By 1918 almost all the villages had been abandoned and the Community Doukhobors under Peter Verigin had departed for a new utopia in British Columbia, while a large number of the original group had accepted individual homesteads.[52] Some of the land that had been set aside as the Doukhobor block was then taken over by the Soldier Settlement Board.

The year 1918 was significant for another communal religious denomination, the Hutterites or Society of Huterian Bretheren. In this year they made their first Canadian settlements, immediately founding nine colonies (or "bruderhofs") in southern Manitoba and twelve near Lethbridge, Alberta.[53] From this point, their role in spreading rural settlement was invigorated by an extremely high birthrate and a regular hiving off of daughter colonies, as well as by a devotion to farming as a way of life. Peters reported that in 1960 over 8,000 Hutterites were in Canada, scattered in colonies of which "Manitoba has over thirty ... Alberta almost twice as many, and Saskatchewan has six."[54] Other employment, such as carpentry or machine repair, carried on by men in the colony, was adjunct to the agricultural livelihood. Hutterite settlement did not approach the boreal fringe during the period of this study, but expansion into a broader range of land types was bound to come as a result of the rapid increase in the number of colonies: once a population of 150 was approached by a colony, preparations began for the establishment of an off-shoot colony on a separate property. Theirs was a form of Christian Communism, in many ways similar to the beliefs of the Doukhobors and the conservative Mennonites, that was singularly successful (though confronted at times by hostility and legislation restricting their expansion). Their lifeworld was their colony.

Two other groups in the western interior had religion-based motivation for settlement. Katz and Lehr document fifteen Jewish and eighteen Mormon planned settlements, the installation of which began in the 1880s, in addition to some individual settlers on homesteads. Some Jews associated with the Edenbridge colony in Saskatchewan had a socialist stimulus for homesteading on the eve of World War I.[55] But, significantly for the study of the late frontier, and perhaps indicating the increasing difficulty of finding sizable blocks of good land, the search for colony sites by these groups petered out by the second decade of the twentieth century.[56]

Most of the refuge-seeking religious groups, except the Hutterites, had been involved in the movement into western Canada by the end of the nineteenth century. At the time of World War I most were well settled, and if some of their members were still engaged in pushing the frontier, it was because their very high birthrates provided a new generation of land seekers. An exception was the Community Doukhobor group which, for external and internal reasons, was in turmoil within a decade after first entry. A slightly later exception was the exodus of many Jews to the more religiously and socially convenient environment of a city, where a concentration of Jewish facilities and people could be found.

The Catholic church was interested in encouraging the settlement of its adherents in newly opening areas across

Canada. As far as group settlement was concerned, the establishment of sizable Catholic colonies in western Canada had been only modestly successful. Church champions of the West were frustrated by what appeared to be general disinterest among potential settlers and leaders in Québec.[57] In the early decades of the twentieth century, the most notable success was the St Peter's Colony of Catholics of German extraction who moved to a large block northeast of Saskatoon, in what would have been part of the northern frontier at the time. A French-language pamphlet describing all the characteristics of central Alberta for intending settlers was published by a priest in Québec in 1914 (with a cover declaration that translates as "The Salvation of the French-Canadian Race – Colonization"), and another in 1928 celebrated the success of French Canadian settlement – "française et catholique" – north of Prince Albert, encapsulating the results of the Church's efforts in "the convents, the schools ... the presbyteries, the churches that adorn our villages."[58] The newspaper that served the Francophone communities in Saskatchewan, *Le Patriote de L'Ouest*, declared on its masthead "Notre Foi! Notre Langue!" (Our Faith! Our Language!).

The dimensions of church involvement and intensity of activity within Québec were larger and rooted in long experience. As shown previously, the Roman Catholic church had a formal role with the government in the 1930s in organizing and guiding new fringe settlement under the Vautrin plan. But the church had never been far from the forefront in what Biays called the "cult" of colonization or, in Morissonneau's version, of creating an ideology of farm settlement as cultural fortress, devoted to maintaining a homeland for Québécois and encouraging expatriates to return to rural parishes.[59] A cadre of priests, called "prêtres colonisateurs," nurtured new settlements both in Québec and elsewhere; and the importance of the church in colonizing is indicated by the position of "missionnaire colonisateur" in the Québec Ministry of Colonization, Mines and Fisheries in 1920, held by the influential M. L'abbé Ivanhoë Caron.[60] Members of the clergy had an indirect but important role through publishing accounts of new settlements and publicizing the opportunity – and in a sense the national duty – of expanding into the margins. Normand Séguin says of l'abbé Caron, for example, that "his works on Abitibi make him the most prolific author – on colonization – of the first two decades of the new [twentieth] century."[61] Such literature carried beyond Québec's borders, and encouraged the expansion of French Canadian settlement in Ontario and the West. Colonization continued to be a favourite topic into the 1940s, when the Church-supported Semaines Sociales du Canada made colonization the subject of its annual congress. The lectures and discussions covered, among other things, the role of the Church, recruitment of colonists, establishing colonies outside Québec, and the imminent demobilization of military personnel.[62]

Many religious organizations were caught up in the *agrarian mystique* and in many parts of the country were active in encouraging settlers to go into the margins. Religious authorities worked with the federal government, especially in the West, in introducing groups of settlers to what were almost always marginal conditions after the first decade of the twentieth century. The Catholic church in Québec continued to be the outstanding example of a religious organization that publicized the desirability of farm colonies and to a large extent inspired and stimulated the provincial government into making internal colonization a century-long initiative.

Examples of collaboration between an immigrant national association and a church, somewhat similar to Québec's church-government arrangement, were to be found in various parts of Canada. In 1930 the Royal Commission on Immigration and Settlement in Saskatchewan was informed of collaboration between the Netherlands Colonization Association and the Dutch Reformed Church. A Mr Henry Hoogeveen reported that:

> The Netherlands Colonization Association ... sends us out a paper every spring to ask us if we can place any boys on the farms or if there are any farms for sale ... These papers go to leading members of the Church in different parts of the West ... Members

of the Church are given the first preference to take [vacancies?] or purchase land … We trust to the Church and its officials to make a wise selection. The advantage of the Association is that it keeps an eye on the immigrant after he comes to Canada … The Canadian National Railways are trying to make a Dutch settlement at Cold Lake, in the extreme northwest of the Province, and we are trying to get 50 families from Holland to settle up there. The idea is that they will homestead … the agent … said it would be better for mixed farming than grain farming.[63]

A wide variety of churches worked closely with both of the major railway systems to introduce settlers, especially in the West.

*Railways*

The railways had compelling reasons for wanting the territories they traversed to be well settled: their lifeblood was primarily the agricultural production of their tributary areas. By the second decade of the twentieth century, the managers of the Canadian Pacific Railway (CPR) were so convinced of the need for a rapid increase of settlers, to make their own land more valuable and to produce freight for the trains, that they vigorously encouraged landseekers to take up homesteads anywhere in the West, whether on company land or not. The second transcontinental railway made its way through the late opened areas of Québec's Abitibi and Ontario's Cochrane District on the eve of World War I, followed shortly by a third that, when completed, would almost mirror it from Québec to British Columbia.[64] Various branch lines were pushed toward points in the boreal fringes of the Prairie Provinces during the 1910s and 20s. The line that became the Northern Alberta Railway entered the Peace River country in 1915, and was close to British Columbia by 1930. The frenzy of railway construction, even toward the marginal areas, indicated that any growing population was of vital interest to the railways; and, at least as profoundly, the railways were of vital importance to the survival of the late settlers. As Robert England phrased it in 1936, "Railway facilities were the main factor in development of a district," and in Québec, a transcontinental railway through what became Abitibi was seen as providing the area with an ocean port on both the Atlantic and Pacific.[65] The highly effective series of maps that appeared with the famous Canadian Frontiers of Settlement volumes showed the invasion of the Prairie Provinces by the railways as tentacles with a belt of access ten miles wide on each side of the tracks. By 1931 these twenty-mile-wide tentacles covered most of the interior plains, even reaching near some of the latest marginal settlement, such as north of Swan River in Manitoba, north of the Saskatchewan River in eastern Saskatchewan, and toward Cold Lake in northeastern Alberta.[66] But, despite great effort, the outcome for both the railways and a large proportion of the population was eventual abandonment, except for a few gratifying discoveries of pockets of good land along – or in a few cases well north of – the boreal edges. People were leaving even before World War II (as most of the population graphs in chapter 2 demonstrate).

Encouragement of potential immigrants by the railways took various forms, including some of the same enticements that had proven successful in populating the prairies in the late nineteenth century. Being first in the field in Canada, the CPR led the way in devising a wide range of schemes to expedite settlement in proximity to its routes. Some inspiration came from models tried earlier in the United States, such as demonstration farms and investment in general agricultural improvements.[67] Although the subsidizing of travel costs of bona fide settlers was a perennial form of assistance, other devices were tried from time to time, and were adapted to particular groups or conditions in Canada or the source countries. By the second decade of the twentieth century the other major railways, especially the Canadian National Railway (CNR) system into which the largest of them metamorphosed beginning in 1919, were developing their own ways of attracting settlers. The "new" CPR thinking – that any settlement was good for business – came to be accepted wisdom in railway planning in the 1920s.

One of the most celebrated early initiatives by the railways was the harvest excursion train that introduced thousands of eastern Canadians, and others, to the farming opportunities of the Prairie Provinces. Very cheap return fares were offered at harvest time to bring in enough field workers to deal with the rapidly increasing acreage of grain. The benefit for the railways was in getting the grain safely onto the trains and to market. But a supplementary benefit was the powerful, first-hand advertising of the wonderful "breadbasket" of the West to many potential farm settlers. The harvest excursionists flooded westwards in the thousands in most years around the turn of the twentieth century, and in the tens of thousands in the 1910s and 1920s, until improved machinery took the place of most manual labour by the end of the 1920s.[68]

The CPR organized a large system of agents in the United States and overseas to provide information and various forms of encouragement to take up land in Canada. The company showed considerable flexibility, offering 20- and even 34-year purchase arrangements, and in some cases postponement of payment on the principal or a discount for payment in cash.[69] By the 1920s it became possible to pay off land purchase with a portion of the crop, and various other modifications of the purchase arrangements were made, especially with the valued group settlements. The CPR's concern to nurture the success of its farm settlers, most of whom came from quite different environments (including eastern Canada), led to a variety of initiatives, such as its demonstration farms and "agricultural trains" that distributed practical farm information and quality seed[70] (along lines also followed by the Saskatchewan government with its Better Farming Trains).

The CPR was particularly interested in attracting settlers from outside Canada, and thus it was good business to become involved as far as possible in assisting immigration.

In this light, it acquired its own transatlantic fleet, including the "Empress" liners, which it continued to operate for decades. This provided its own integrated system from European ports to the Canadian western interior, and streamlined the arrangements it could make with religious or national groups, such as the Lutheran Immigration Board, the German Catholic Board, the Scottish Immigrant Aid Society, or the Mennonites fleeing the Ukraine in the 1920s.[71] It also led to collaboration with the Canadian government, when both realized a common interest in rapidly filling the remaining empty land of Canada. This interest converged in a meeting in Ottawa in 1918, and eventually evolved into the Railway Agreement in 1925, in which the railways would take the lead in recruiting and transporting immigrants while government set the regulations.[72]

The agreement of 1925 involved the CPR and the second transcontinental railway, the recently consolidated CNR.[73] The CNR tried to attract settlers in many ways similar to those of the CPR, including continuing certain of the sirenic activities of the precursors from which the national system was formed. The main interest of both government and railways was to fill in the land; to avoid unsuitable urban applicants, the immigrants were to be restricted to "agriculturalists, agricultural workers, or domestic servants." The CNR, through the policy statement of its new Department of Colonization, Agriculture, and Natural Resources, made clear that its goal at this time was to secure agricultural settlers for lands adjacent to its rails. The CNR established various kinds of contact persons overseas, and made alliances with a number of independent steamship companies to transport immigrants. Modern tools, such as radio and film, as well as the more traditional newspaper advertisements, posters, and lectures, were part of the railway's arsenal of enticements (see illustration 4.1).[74] The railways beckoned immigrants, as evidenced by Hoogeveen's testimony to the Saskatchewan Royal Commission on Immigration and Settlement (above), to the far reaches of their networks and of the Canadian ecumene in the inter-war

# The Great Clay Belt of Northern Ontario

## Twenty Million Acres of Virgin Soil Await the Farmer's Plow and Reaper.

### In New Ontario's Great Clay Belt All Can Strike it Rich. Available Farming Lands are Being Rapidly Settled.

**Where men with Determination, Good Health and Strength need have No Fear of Failure. Forest Line Rapidly Receding from Onslaught of Settler's Axe. Experiences of Settlers Who Have Made Good, as Related, by Themselves. Some of Their Hardships in Days of the First Settlement Before Building of Government Railways and Construction of Colonization Roads. The Life of the Average Settler and the Many Advantages the Natural Conditions of the District Afford as Compared With the West. Timber for Building and Nearness to Market Are Big Advantages.**

New Ontario—the name sounds familiar, quite familiar indeed, for one can scarcely pick up a newspaper that does not contain news of some kind or another from that district. Yet, well known as is New Ontario, how many have any idea of the vastness of this new territory, with its wealth of timber, its healthful climate, and a soil capable of producing grain and vegetable crops surpassed nowhere in agricultural Canada? It is the heritage for the man with ambition and courage who wants to shuffle off the yoke of the wage-earner, the under man, and gain a living of independence. One must actually visit New Ontario before he can form any idea of its possibilities. He must go there to see for himself the great opportunities that exist.

**Desirable Settlers.**

In all of the more recently settled districts of all Canada it is doubtful if in any one section of this country can a more desirable class of settlers be found. About ninety per cent. of

5

Illustration 4.1
A railway company glorification of the boreal frontier in New (i.e. Northern) Ontario.
*Source*: First text page of *The Great Clay Belt of Northern Ontario*, pamphlet No. 11, Temiskaming and Northern Ontario Railway Commission, 1912, courtesy of Archives of Ontario (PAMPH 1912 #27).

period. In addition, the CNR had people "on the ground" in settlement areas to help landseekers purchase and locate. One such agent (Androchowiez, see above) claimed to have settled thirty-four families north of Prince Albert in three years. His success underlined Sifton's opinion of the usefulness of eastern Europeans in settling land that North Americans would disdain:

> The land in question, with the exception of six parcels I would say was totally abandoned by people who moved into the cities. This land had been occupied and homesteaded some years ago, but left unattended, growing into weeds, shacks abandoned, and things like that. As to why they left … Possibly it was because it was too tough a country for them … The people were mostly French, a few Scandinavians, and a few Canadians who left these lands. I also settled 38 families, at least I assisted … north east of this city … The districts were Henribourg, Paddockwood, Meath Park, Foxwood, and Samburg … The countries from which most of these 72 families came are Poland and Ruthenia … I have them sprinkled amongst the English speaking people … and while some of them objected I find they pick up the ideas of this country much quicker than when in groups … In the winter time they cut cordwood, pile it up, season it, and then bring it into town and trade for groceries.[75]

## THE FADING OF A POWERFUL IDEA

The magic of the New World had been based on the opportunity to own land, in contrast to an Old World where ownership of real property had been in relatively few hands. And so for over two centuries the odyssey of the landseeker had advanced across the continent, gobbling up almost all unalienated land.[76] The stimulus was a yearning for livelihood opportunity on the land, expressed in French Canada as "le patrimoine" or later "le mythe du Nord," and in the east-west movement across the continent as "manifest destiny" (with additional nationalistic overtones) or simply The Frontier. An odyssey is at heart a search, and the search for farmland had gone on unabated for generations, leading to a continuing momentum of the idea of the frontier and a yearning for its extension. But by the beginning of the twentieth century there was a growing premonition that the supply of good farmland in the New World countries could not last much longer. Apart from the scholars who had been alerted by the Superintendent of the United States census of 1890, the general public was being told, as the *Century Magazine* worded it, "we are within sight of the end of free land ... the last West ... has been reached."[77] It is not surprising that the search for land took on a touch of desperation by the end of World War I, and that a society fashioned by frontier opportunities would make great efforts to perpetuate the legendary era of expansion. The efforts were contributed by individual and group settlers in invading the margins beyond previous settlement, but as much by persons informed by science and the broader picture, such as politicians, bureaucrats, and religious leaders, who had less innocent reasons for encouraging what were questionable settlement experiments. As Morris Zaslow explains, for the latter persons "That farming should be extended as far as natural and economic conditions permitted was an article of faith" in keeping with their belief that "high rates of immigration and frontier settlement were the sure recipe for national success and greatness."[78] In the minds of politicians and bureaucrats farm settlers were of infinitely more value to the nation than raw land. Expansion of the ecumene was a matter of national pride.

# 5

# Living the Marginal Experience, From Abitibi to Peace River

Everybody has a log house ... poplar logs ... plugged up with green moss ... We just have a dirt roof ... I often wonder why we worked so hard, for the rewards were so pitifully small. This is such a narrow life in so many ways.[1]

You could climb to the top of a tree and just get a very faint idea of what a place must be like and that's all. So, it was just a gamble.[2]

This chapter attempts to edge closer to the actual experiences people weathered on the inter-war margins. Living conditions in the areas in Canada struck by drought have been well publicized and illustrated, as in the United States and distant Australia, especially through the arresting imagery of the Dust Bowl (see illustration 3.4).[3] This book draws attention to the opposite of the dry margin, namely the boreal margin that was promoted as an oasis for Dust Bowl refugees: the northern woodland at least could provide the precious moisture the arid belt lacked. But this was not an unqualified alternative: the boreal fringe was in its own ways seriously marginal. For one thing, the boreal expansion was back into the woodland that had been left behind in the advance onto the prairies nearly a century earlier in central North America,[4] and the northern forest was a comparatively impoverished example of poplars and spindly evergreens.

The chapter opens with some accounts by or about individuals or settler families who tried out the boreal fringe in locations beyond the favoured parkland and into the bush. Here the spotlight is on the individual, and a fundamental aspect of the frontier lifeworld, whether one entered as a single landseeker, or with a family, or as a member of an association, was that living in a marginal area was *experienced uniquely and individually,* despite the family or group experience of which all members partook. There could be family success or failure, or the completion of a group community hall, or collaboration to fight a fire, but this was not the same as the personal perceptions, coloured by private emotions, of the day-to-day struggle, some of which might surface in recollections. At this distance it is hardly possible to recapture the full intensity of daily happenings; but much meaningful evidence comes through in the accounts or tangentially in the decisions reported. In moving from records of individuals through those of group settlements it should be possible to see that some experiences were common to all people who tried *to make a go of it* in conditions marginal for farming.

Migration to the frontier was seldom a purely individual odyssey. Many of the celebrated "lone pioneers" were going ahead to seek out a location for other members of the family and knew that the others would follow in a few months. In other cases landseekers were following steps taken earlier by relatives or friends, then going a little farther. This so-called "chain migration," while originating with a precursor who was likely a relative, did not extend in a straight line. It was more a line of inspiration, and followers could swing through wide arcs while attempting to maintain ties or a measure of propinquity.[5] The migration to frontier land by a family or a group was chain migration at one extreme, with many "links" of the chain clustering together. It was as common in the late stages of the frontier, as illustrated by a number of the cases below, as it had been in the nineteenth century.

That there were bachelors, too, cannot be ignored, but only a small percentage of them remained bachelors for any length of time: most found a wife before long, while those who failed to attract a mate either became accepted as "the local bachelor" (which usually contained a modicum of pity) or moved away.[6] The bachelor was caught up as a committed male homesteader in what Danysk discusses as "the construction of gender." A sharp distinction was drawn between men and women (as identified in chapter 2). Speaking of the persona of the bachelor (who was usually thought of as a homesteader heading for marriage), Danysk identifies physical prowess (strength, stamina, dexterity), individualism, a spirit of adventure, courage, and resourcefulness as ideal characteristics. Men on the frontier, whatever their social or marital status, were typically striving "for full economic membership in the agrarian community," and farm failure or interminable unproductive labour could deal a blow to self esteem.[7] The resident bachelors often turned out to be the odd individuals, the "characters," on whom recollections tend to focus. Peculiarity was probably no more common on the frontier than in a city, but with the low density of population, the unusual person seemed to loom larger and to take on legendary characteristics more readily. A classic example was the Peace River country's "Twelve-Foot" Davis. Actually he was short and muscular, but acquired his odd moniker in the Cariboo gold rush where he discovered a twelve-foot-wide error in earlier claims and mined a small fortune from his narrow strip. His real fame, however, arose from his strength, good nature, and friendliness as a settler near Peace River Crossing, Alberta.[8] In the light of the many unusual personalities and exclusive groups, the frontier seems to have been a *habitat of otherness*, and certainly for both individuals and groups the isolation allowed for non-conforming ways of life.

The experiences reported in this chapter take a variety of forms. Some are written reminiscences, others are products of questionnaires or interviews (some tape-recorded), and still others are extracted from third-person reports. They have been found in archives or family collections, or have been discovered and captured through field work. Some are more formal proceedings of government commissions or extracts from government agency reports. Most have never been published, but the few that have been available in some public form are used here because their uniqueness or significance warrant more exposure. The provenance and context is explained with each item. Communication seems to have shown a gender distinction in the pioneering population. In the early days of the frontier in North America, letters "written home" were often penned by women, even though signed by the male head of household. The writing by women was carried into the early twentieth century, when a large proportion of the letters recording the deep feelings of settling new land in Canada were by women. Note the women's voices, usually providing factual memories, in the backgrounds of the interviews of their husbands in this chapter. In contrast, an overwhelming proportion of those that testified before commissions of inquiry or conferences or who published research on the marginal settlement were men. This chapter amalgamates the communications by men and women by drawing on a variety of sources.

Source materials that survive for the better part of a century might be suspected of being somehow privileged and not typical. All of the stories presented in this chapter, however, are of people who attempted to make a farm in a boreal location (excepting Dr Mary Percy). In terms of the social levelling and simplification of the frontier and the egalitarian impact of similar marginal natural conditions, these people can be seen as broadly "typical." In one sense they are exceptional: of all the people who at one time or other resided in or moved through the boreal margin, they are among the few whose stories survive.

## LONE VOICES: INDIVIDUALS AND FAMILIES

It is rather uncommon for a reminiscence to be entirely self-centred: even though it may be one person's story, there are usually asides indicating involvement of others either in the same household or vicinity, or somewhere along the migration chain. A lone voice, even from one who genuinely lived alone, does not have exclusive claim to loneliness on the frontier. It was not unusual for rural wives, isolated all day and even longer around the homestead, to suffer fear and loneliness sometimes to the extreme of "cabin fever." Hilda Rose, from far north on the Peace River, realized that "it takes some mental calibre to come in here and live alone and not see a white woman more than twice a year. If you haven't got much in your head the lonesomeness will get you."[9] The men, including the bachelors, were out and about meeting other people and more often experiencing different environments. And although being alone was rare in group settlement, sometimes for an individual the group itself could be a negative environment.

### *Abitibi*

The landseeker in Québec in the first few decades of the twentieth century was informed by an 80-page booklet called *Le Livre du Colon* (*The Book of the Colonist*, with a subtitle offering "how to establish oneself on land for next to nothing"). It is a compendium of useful knowledge, ranging from choosing a lot to clearing bush; from crops to try, to harvesting wilderness products, to treatment of health problems of humans and livestock, and many tricks of backwoods survival.[10] Under the heading "The way to assess the quality and nature of the soil," it claims that the goodness of the soil is known by the vigorous growth and the cleanness of the bark of the trees. Black or nearly black soils, that give the same colour to water that lies a while on the surface, are of good quality. It goes on to advise about types of trees, which of course would have been fundamental knowledge for any part of the boreal fringes: for example, mixed woods with "maple, wild cherry, a little ash, basswood and spruce, indicate generally good land"; and this is accompanied by a listing of twenty-two trees with the kind of land they prefer.[11] For the uninformed landseeker, however, including many in the back-to-the-land movement, an immediate problem would be identifying trees by type followed by understanding how to apply the identification in the rudimentary farming effort.

It is not known whether Joseph Laliberté availed himself of *Le Livre du Colon* when he pioneered in the western extremity of Abitibi in the 1930s. He probably had more than average knowledge, having taken an agricultural course at the college in Ste Anne de la Pocatière, by the St Lawrence River, and he was aware of benefits available to the colonist. He had his sights set on a brand new frontier area and, with a healthy ambition to make himself a man of importance, he undertook the tortuous trip by motorboat from La Sarre, on the transcontinental railway line, to roadless Roquemaure on the south side of Lake Abitibi (figure 5.1). He had been in the region in previous years, having two brothers elsewhere in Abitibi. He might be thought of as somewhat untypical in having had a fairly relevant agriculture course, but not in having some links of his family migration chain already in the area. Laliberté chose his location and entered as a colonist in mid-May 1935.

The Canada Land Inventory assessment of the Roquemaure area for agriculture, in figure 5.1, shows the crazy-quilt variability in the land that was typical of most of the boreal forest settlements across the country (see further explanation of the Canada Land Inventory symbols in appendix B). The lateness of opening this territory south of Baie La Sarre is explained by the quality of the land: most of the area is classed as lower quality than class 3, thus almost ruling it out as crop land, and even land that might be used for hay or pasture is often limited by poor drainage (subscript W) or "adverse topography" (T). Much of the area is shown, by the hatch marks, to have a complex surface, thus suggesting difficulty for agriculture. The section to the northeast around Palmarolle and Poularies and north to Macamic on the transcontinental railway provides much more class 3 land, but almost everywhere drainage is unsatisfactory, permeability is low (D), and here and there soil is shallow

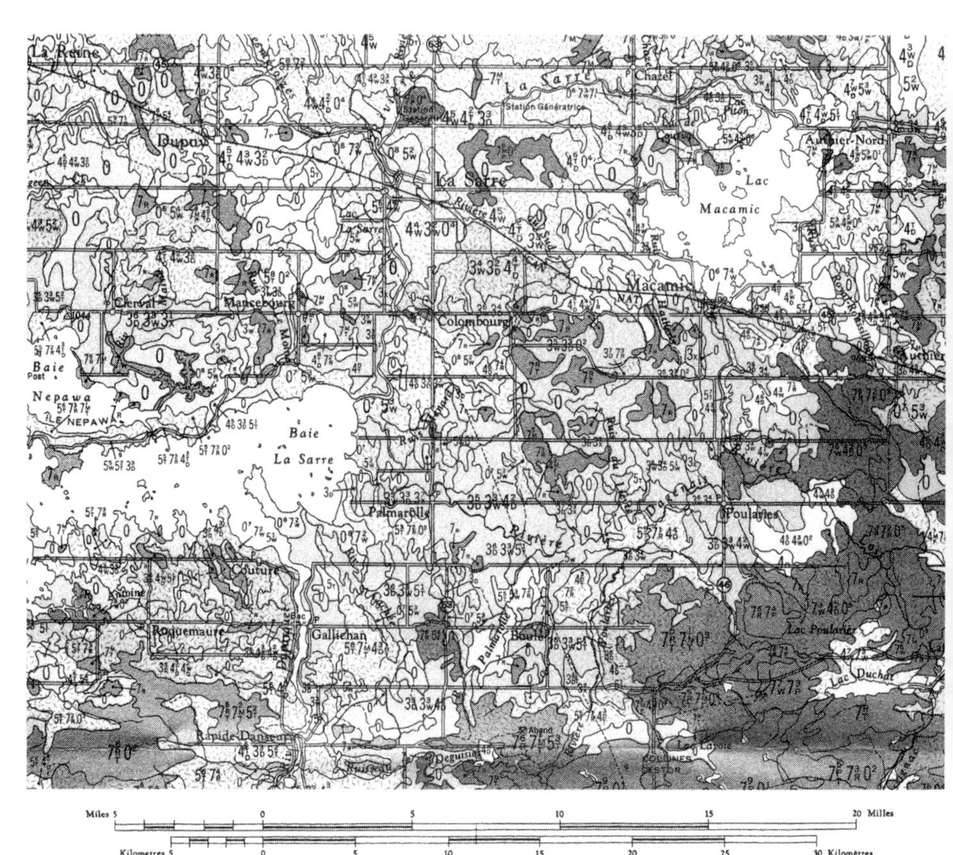


Figure 5.1
Black and white photographic rendition of the Canada Land Inventory [hereafter CLI] assessment of soil quality for agriculture, Abitibi District, Québec. Classes 1 to 3 soils (the larger numbers) are considered suitable for plough agriculture, with class 4 useful for pasture or hay. Classes 5 and 6 may yield rough forage, but Class 7 ("rockland") and 0 (organic, such as bog) are non-agricultural. Regarding the search for farmland, note the excessive mixture of soil conditions (the speckling indicates complexity), much of it poor quality and often blocking access to usable land. The smaller numbers indicate how many tenths of the area are occupied by each class: e.g., much of the land along the road through Roquemaure's eastern neighbour, Palmarolle (centre), is class 3, but immediately south is an area that is 3/10ths class 5, 2/10ths class 7, and 5/10ths organic, thus unlikely to be of any use for agriculture. For further clarification of the CLI, see appendix B.
*Source:* Part of CLI map 32D, E "Noranda-Rouyn."

(R). The climate would be similar to that for the eastern Cochrane District, discussed in chapter III, where summer frost is a possibility and dull weather limits the build-up of heat units needed for valuable crops. The overall picture, especially for the Roquemaure district, is of unequivocal marginality for agriculture.

Joseph Laliberté was intending to take land, as well as to serve as the parish agronomist. His reminiscence, while not representative of the uninformed settler, gives a good impression of the range and dimensions of the challenges confronting any individual colonist entering new land. From the landing place, he was conducted to the priest's house by a local boy:

my belongings tied to two logs fastened at one end to the front of the wagon, the other end trailing in the mud and water. In the low spots, we walked in more than a foot of water. We were at the end of the melting of spring snow and the road was … very rudimentary.

My arrival … did not impress the priest … far from it. I weighed 120 pounds and I looked like a high school kid on holidays … I bought the lot reserved for the agronomist for $30. This lot had been ravaged by fire 8 or 10 years previously, at the time of a very extensive conflagration … My lot was sprouting up with young 10- to 12-foot aspen. I found there a cabin of 16 by 16 feet, built the year before by Mr Charest who had just left. The floor of my little cabin was of logs roughly squared, so that the dandelions pushed up between the logs. There was a table of rough boards nailed to the wall and a bed made of the same material. In keeping with the rest, the roof was of poles, earth and a thin roofing paper. In reality this was not the Chateau Frontenac …

I had never done logging … never worked in forestry … I bought an axe head, a box saw and various accessories. I returned to my place, put a handle on my axe and sharpened it. And *voilà*, I was ready to attack.

When I arrived on my lot, there was already half the road cleared opposite the rang where my lot was situated. The incoming settlers had been clearing the other half and ploughing a ditch on one side. So I prepared myself to fell a large, dried birch of pretty nearly 20 inches diameter at the stump. That took me half a day …

Fortunately, one of the neighbour ladies agreed to sell me some bread, and my mother had put sugar and tea in my baggage. No one was able to take me as a boarder because food reserves were exhausted in all the families, since the boat … was late … being repaired at La Sarre … I had to fast for 15 days. The menu was composed, for three meals a day, of bread, drippings, sugared tea and a little hare from time to time.

… during the winter 1935–36 … a saw mill built by the Ministry of Colonization [was] put into operation … the Vautrin plan was gradually applied … During the winter people had cut wood and transported it to the mill yard because in summer the roads were not passable.[12]

Even a number of years after opening for settlement all settlers suffered many hardships. Joseph Laliberté struggled on and eventually did become "a man of importance," at least in the district.

### *Cochrane*

In the adjacent part of Ontario, a few dozen miles west of Joseph Laliberté in the so-called clay belts, people were seeking suitable land for farming (see illustration 5.1). The possibility of finding usable land was only slightly better than in Roquemaure. Along the transcontinental railway west of the town of Cochrane the land is generally a mixture of classes 3, 4, and 5, with some large stretches of organic soil (O; figure 5.2). Areas with complex land surface, providing difficulties for using farm equipment, are mainly away from the railway. Throughout the region a major limitation for agriculture is climate (subscript C), primarily the threat of summer frost and commonly a dull autumn (see chapter 3), and problems with excess water are ubiquitous, as Fernow observed in 1912.

With the inauguration of the Gordon plan a number of municipalities throughout Ontario were encouraging their

Illustration 5.1
Settler with ox cart c. 1914 [Cochrane, Ontario]. Oxen were vitally important draught animals in pioneering throughout the nineteeth century and into the 1920s.
*Source*: Archives of Ontario, RG1–448.3 Box A #163.

welfare recipients to participate in the back-to-the-land movement even though many of them had no farming experience. With a $600 enticement, to cover costs of getting established and buying supplies for a couple of years, the city welfare people found it hard to refuse. The result was a widely publicized bad name for northern Ontario, but at the same time an attempt to lay blame on the unsuccessful settlers and the officials who sent them. The Crown land agent at Matheson had no doubts:

The trouble with these people is that they aren't adaptable to farming and homesteading in the north … none of them should have come here. They seemed to look on the jaunt north as a picnic. None of them has cleared a foot of land since they've been up here, and I am under the impression that they seemed to think

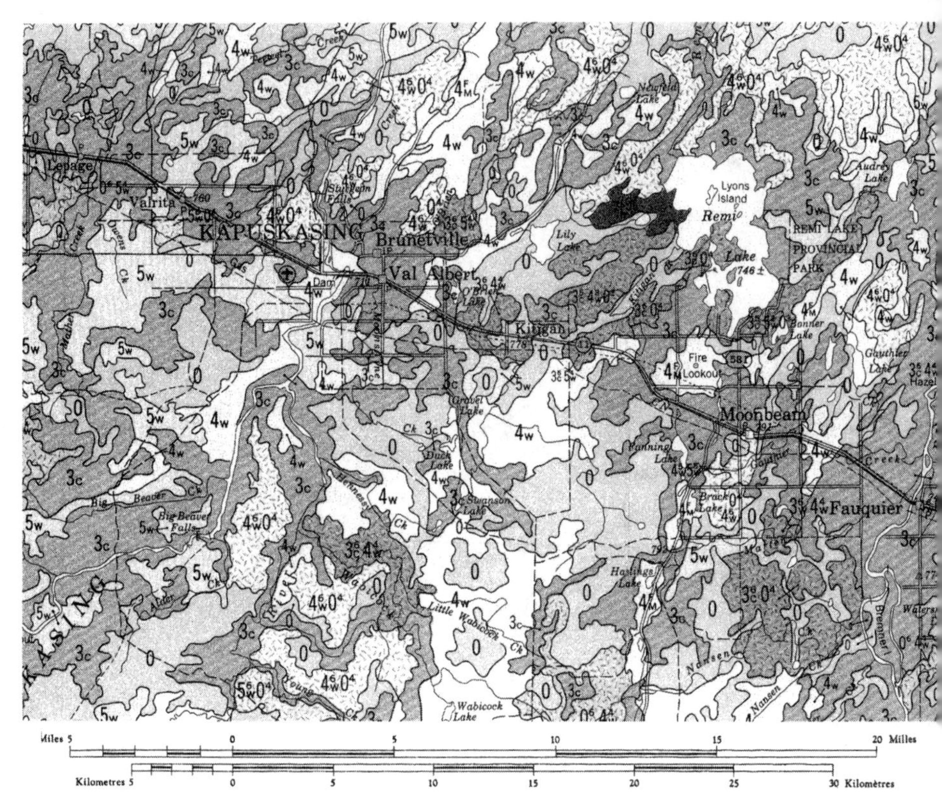


Figure 5.2
CLI assessment of soil quality for agriculture, Cochrane District, Ontario. Note excessive variability at odds with the supposed homogeneous Clay Belt. Each CLI map is a black and white photographic copy of the full-colour published version. Also see appendix B for the CLI.
*Source:* Part of CLI map 42G "Kapuskasing."

that since St. Catharines sent them here all that they had to do was take things easy and … St. Catharines would care for them indefinitely … These men … were unfamiliar with bush work, and they wouldn't clear any land.[13]

There were cases of what has been called *The Wretched of Canada*, most prominent during the thirties when people wrote pleas to the Prime Minister saying "My children has not enough to eat. Please try & do something," and " I am a girl thirteen years old and I have to go to school every day and its very cold now already and I haven't got a coat."[14] There were obviously cases of genuine hardship, even for some who were willing to work. The blame was not all on one side. As Gosselin and Boucher conclude in their valuable study of the administration of settlement in the twenties and thirties, "In Ontario, apart from the sale of farm lots to settlers at low prices, and the construction of colonization roads, very little has been done to encourage and assist land settlement or to induce settlers to clear and improve their farm lots and become self-sustaining. In Québec a serious attempt has been made during the last decade to make land settlement more successful … in the form of grants for building houses and barns, and for the purchase of equipment and livestock. In order to induce them to clear more land and raise more crops, substantial land clearing bonuses are given … In addition, close supervision and advice on farming methods are given … The Church also plays an important role."[15] This provincial contrast is highlighted by Joseph Laliberté's story of success in comparable conditions. Some success stories on the Ontario side emerged during the back-to-the-land debate. One settler near Cochrane in the Great Clay Belt wrote:

> The winter is very cold and hard for all of us but we couldn't expect anything else for our first winter. It is rather hard to live on ten dollars a month though and there have been times when we have run pretty short, but never to the point of starving …
>
> We had a doctor out from Cochrane to me but not because I wasn't getting enough food but because I had another attack of pleurisy from which I suffered all last summer. The air is better than any medicine for I had a terrible cough when we came and now it is nearly all gone.
>
> … if a person has enough grit and backbone they can make good if they try. It will be mighty hard for the first year or two but a person has their own home and that is something.[16]

*Battle River Prairie*

From the opposite end of the boreal margin, in the northern Peace River country of Alberta, comes what one might think of as a relatively objective description of experiences faced by some of the trailblazers. Mary Percy was a medical doctor who wrote a large number of letters to family members in England from 1929 to 1931.[17] During those years settlers were continuing to flood in to take up uncleared land in the district then called Battle River Prairie (around today's Notikewin and Manning). The choice area for settlement was the grassland and timber (mainly aspen poplar) mixture called parkland, but not far away it graded into the more typical boreal forest. The newcomers included many non-English speakers whom the young English doctor came to respect for their commitment to hard work and their ability to withstand severe deprivation. Although Mary Percy was essentially an observer and correspondent, and although the hardships were not directly hers and the descriptions might have been coloured a little, they no doubt come close to the truth. Dr Percy went by horseback for a second time to a maternity case ten miles from her cabin:

> but the trail had drifted over badly and it was very slow, very hard going … It certainly was the slowest and the most tiring ride I have done … I was there Thursday for 12 hours. The people are Ukrainians, speak no English. There was only a wooden bench to sit on … When the baby was born, we had to wrap it in a dirty old cloth skirt. They hadn't so much as a binder, napkin, blanket, or clean rag! … Poor little beggar – his skin was chafed …
>
> My old pal Kolyna with the heart failure is back from Peace River … When they got there they found the hospital closed … So they brought him all the way back!! That's 200 miles in an open sleigh for a man who is blue-green and swollen all over with heart failure! He ought to have died, but of course he hasn't. They're awfully tough.

The amount of TB up here is horrifying. I think I told you about the [Métis] family ... It's simply heart-breaking. And during the last week I've also collected two new cases amongst the white population – a man of 31, dying fast, and a doubtful case, a girl of 6 who was living in the same house as the young Norwegian boy I sent out to a sanatorium just before Xmas. I'm also looking after two more certain cases and three more doubtful cases and the population of the district is only 1,200.[18]

Dr Percy's narrative, although emanating from a somewhat untypical and privileged view of a corner of Canada's boreal margin, has the ring of being written "on the spot." She was herself to become rooted in the region, marrying a farmer and spending the rest of her life there.

Dr Percy's account is augmented by another well-informed "outside" view, provided by the annual report of the Alberta Provincial Police in the Peace River subdistrict. In the General Remarks submitted in 1930 (apparently for the year 1929), it reported that:

The past year has been a very busy one both in the enforcement of the law as well as with the attention given to destitutes. There has been a very large number of new settlers come into this part of the country during the last twelve months, especially so in the Battle River district, Hines Creek and the district around Falher and McLennan and High Prairie. As is to be expected, many came in without any means of carrying on and after filing were practically destitute. I find the majority of them came in from Saskatchewan and the eastern part of the Dominion, some who are too old to face the hardship of pioneer life. Very few I find seem to get right down to the old style of breaking up the land themselves. They go out to work for a few dollars and hire their breaking and seeding done for them. I have not seen one team of oxen at work since being here, tractors yes, but unfortunately not paid for. I find there is around 3,000 homesteads taken up again during the past year. I think that the matter of homesteading should be taken up, now that it is under the Province, with a view

Illustration 5.2
First seeding, near Hines Creek, North Peace River district, Alberta.
*Source:* Provincial Archives of Alberta, photograph A5915.

to discouraging the idea that all is necessary, is to file and pay $10.00, and then demand work and keep ... Some work was obtained during the summer by those willing, on the extension of the railway from Fairview to Hines Creek ... The crops throughout the Sub/District ... were good, but unfortunately the price of grain, took a slide and left the farmers flat, necessitating relief.[19]

Aspects of the northern Peace River pioneer scene, specifically near Hines Creek (30 km northwest of Whitelaw, figure 1.2), are recorded in illustrations 5.2 and 5.3.

Illustration 5.3
Settler's cabin near Hines Creek.
*Source:* Provincial Archives of Alberta, photograph A5921.

## *Southwestern Peace River*

At a slightly earlier date in the settlement of the Peace River country, but in conditions similar to Dr Percy's district around Battle River Prairie, a German-American couple moved to a homestead near the outer edge of surveyed land, close to the outer northern edge of the aspen poplar parkland surrounding Grande Prairie, in the approaches to the Saddle Hills. The area is mainly in classes 2 and 3 for agriculture, with climate and some rough topography as limitations. La Glace can be located near the centre of figure 5.6, and in figure 2.1. The couple's memories of homesteading forty kilometres northeast of Grande Prairie were captured on a tape recorder by a field researcher. The version of the interview offered here is a product of excerpting and editing about a third of the original without changing the order in which the information was given. Some of the flavour of the interview is lost by reducing pauses, repetition, and some phraseology, and especially by not attempting to reproduce the wife's German accent. The husband was twenty-one when the couple came into the Peace River country in 1912, from Wisconsin, following his brother. In the interview many years later the couple laughs about the arduous trek into the Peace River country, with horses and cattle, being their honeymoon. Their start was almost from "scratch," because the husband was not a seasoned farmer. The frontier conditions were made harder in the early years when neighbours abandoned and left land vacant, thus allowing the proliferation of insect and animal pests. This couple stayed on their land because, having put everything they had into the initial entry, they did not have sufficient resources to move anywhere else. A question by the interviewer is shown by a Q, followed by the Kinderwaters' answer.

Q: What persuaded you to move … to the Peace River country? Well, the Canadian government had posters all over the place … "Go West Young Man" … and "you can travel all through Canada for one cent a mile." My brother was up here … he persuaded us to do the same thing – a homestead and a South African scrip.[20] You could buy them in them days.
Q: What kind of a trip did you have here? Oh, we come by the way of Edson.
Q: The Edson Trail was open then? Well [laughter] open … a trail through the bush, that's all. [Mrs Kinderwater: In the summer time it was worse … Well, you see, we had cattle and we had to come in summer so they could feed along the road.]
Q: Did the land look better around La Glace? No, it was just a matter of water mostly … [Mrs K.: Well, we had cattle so we had to be near a crick [creek].]
Q: Do you remember when you bought your first tractor? In them hard thirties we didn't have any tractor. [Mrs K.: … we lost our horses in the fire … and then we still bought horses … and that was '28 when we had the fire that burned the barn.][21]
Q: What was your first crop? Well, the first one was oats, of course. First. We had to have feed, y'see. And then we had a quarter section leased right next to us … from the government. Well, that was hay land, and we put up hay there for our stock.
Q: When did you start growing wheat? Oh, that wouldn't be … I imagine about five years after we came. Do you remember Ma? … about five years after we came.

Q: What did you do for money? Were you self-sufficient from the farm? We always did milk cows, y'know. 'Course we always had meat of our own, and chickens as well, eggs … So we didn't have to buy very much … excepting sugar … We used our own flour, too. [Mrs K.: We had our own cream of wheat. There was a mill in Sexsmith. Then we had our flour for the whole year, because it aged, you know. And by being aged, it got better.] …when it got aged it got dried, and it was better flour. [Mrs K.: We figured we had to have vegetables in order to live. Well, sometimes when the potatoes froze too much we had to go to rice and beans, too. But seldom. It never froze so directly that we had nothing at all.]
Q: Mr K., how old were you when you first came up? Twenty-one. [Mrs K.: And we stayed here twenty-five years before we could take the first trip out, back to the States.]
Q: What did you do for entertainment? [Mrs K.: Oh, we had … the schoolhouse would be the local centre for entertainment. School picnics … and get together with the neighbours. Seems somehow that everybody knew the neighbours better than we do today … and maybe somebody would be puttin' on something for the children, then we'd gather in the schoolhouse. Play cards, or just a social evening.
Q: When you were hauling, how long did it take you to make the trip? I made ten trips over the Edson Trail, two trips one winter. In summertime it didn't take us so long. How much did it take us, Ma? [Mrs K.: Seventeen days.] We had a good team of horses … one way, from Edson here.
Q: How did the "Dirty Thirties" hit you? Oh, well, same as everybody else. [Mrs. K: The grasshopper years were worse than the frost.] Oh, yeah, we had 'em for three years. [Mrs K.: By that time so many of the bachelors have left the country and there was really few people. So to poison all the vacant land … we couldn't do that, so we done what we could. Just when harvest time would be ready they would come like a cloud … They took everything in the garden except the rhubarb … and the potato tops. But everything else green would be stripped in one night. Three years, yeah. But not all the prairie had it, only on the fringes … the grasshoppers came in '25, '26, then we got neighbours.] Y'see, they hatched on that vacant land.[22]

*Northeast Saskatchewan: Meath Park*

Five years after the grasshoppers invaded the Kinderwaters, and about the time that Dr Percy's letters were coming to an end, a young Saskatchewan-born farmer chased off "the bald-headed prairie" by drought was attempting to start a farm in the central section of the boreal fringe. His name was Gabriel Kerpan. He and his parents were representative of the thousands of farmers leaving the prairie for the boreal margin because of worsening drought crises in the 1920s and 30s. His parents had come from a peasant society in southeastern Europe, where their first child was born. The husband had worked in the United States for a few years before he brought his family to settle in Canada. They farmed on the Saskatchewan prairie for twenty-five years before moving to the boreal margin sixty kilometres northeast of Prince Albert (figure 2.1). Gabriel had gone north to help his father clear new land, and chose for himself a largely uncleared quarter section (160 acres) next to his father's half section. When Gabriel chose his land other settlers were already in the area and many, including some relatives, were entering especially to the east, but few were venturing more than a couple of kilometres further north. He married, had four children, and cleared his quarter section in the decade after settling.

Almost sixty years later Gabriel Kerpan recounted the pioneering experiences in response to questions (although edited for clarity, for an authentic "feel" much of the turn of phrase has been retained):

Q: What proportion of your land was wooded? It was all wooded. There was ten acres on the west end … that was ploughed and that was awful sandy. The rest of it was rich loam.
Q: Did you have any government assistance? No … and no loan. There was no advice on the land or anything. I got the seed from my father.
Q: What crops did you plant to start with? Oh, I planted wheat and some oats … because I had summer fallowed that ten acres.

The first year I ploughed, broke, about forty-some acres, forty-eight. And the crop off of it was about sixty-five bushels to the acre, and when we left southern Saskatchewan you couldn't hardly get your seed back. And it was quite wet up there.
Q: Was there a special plough for breaking? Oh, yeah, it has a blade in front of the share … about an inch by six iron. It's drawn out and sharpened and it just slices through the roots and everything. You couldn't plough the land any other way. And, the weather, there was a lot of frost, and early frost. The winters were awful long.
Q: The year you took the land was …? 1934, and we were married in '36.
Q: Did you have drought problems there? Well, in '38–9 it was quite dry. But that land was so light you had to summer fallow it, and you'd get a monstrous crop every time. We had a real good crop the year we sold out and moved to B.C., but you couldn't sell it. I made a deal and some guy took it on a debt … There was no power, no telephone, and you just … when you wanted to contact somebody, it was mocassin telegraph. I think it was the first year we were married there was a fire started down at the tracks in the muskeg, and that was awful. It just created a wind, and we fought that fire … threshed all day … and then at night we fought fire. It burned down about eight feet in places. It burned for months.
Q: For food, did you butcher any calves? [Mrs Kerpan: We didn't ever raise that many and we didn't sell them. To have cows producing you had to. We went to this old bachelor's place to buy a cow and we got it for twelve dollars, wasn't it? And we took her home behind the sleigh box, but it was so cold we had to walk home behind to keep warm. That's where the old man made his own bread. He had it on the kitchen table rising in lard pails, and all the chickens were in the house to keep warm. The chickens were all up on the table picking at the bread as fast as it was rising.] But, that sleigh … I got some poplar boards and shaped them out, and started cutting out the sleighs – runners – for a bobsleigh, 'cause we had no mode of transportation. And in the house I drilled all that iron by hand and the sleigh shoes and everything, bolted them on.

Q: Any livestock failures? [Mrs K.: The only thing we ever lost to illness of our stock was Dan, the standardbred's colt. She was bred to a Percheron, and he got sleeping sickness, encephalitis.]
Q: You didn't have any vaccines then? No, no, they just told us to grease their ears with axle grease so the flies wouldn't get in. They were concerned about the stock getting it, because stock meant a lot. Anyhow, he had to be shot, and we burned him in a straw pile.
Q: Where did you go to church, if you went? We went to a French settlement by the name of Albertville, which was about twelve miles away, or more. That was a major trip.
Q: Were there any social organizations or support? No, there was nothing. Everybody done their own and there was gatherings. If you didn't have machinery, why, who had it, you used it; and if you had anything, they used it. The blacksmith we didn't pay much cash to him for what blacksmithing he done for us because, hell, in the fall of the year we'd haul grain to Prince Albert and get it gristed and make flour … We'd take a bag of flour and a bag of potatoes. And when we'd butcher we'd take a bunch of meat …
Q: What about water supply? Ho, ho! There was an old Syrian in Prince Albert that could go in a trance, and he … couldn't see anything, but he'd take a pencil and he'd draw a map of the farm … where the buildings were, and he'd put a mark and that's where you'd get the water. So we went home and I dug by hand. Went down about twenty feet, and I went back to him and said "There's no water there." He said, "Yep, there's water. You just go a little more." There was a man that had a post hole auger … and pipe connections on it. I went down about fifteen or twenty feet, and when I struck water it shot up in the bottom of the well. I just got out …
Q: Your first house was the granary? No. The first house we lived in was sixteen by sixteen [feet], but it was sitting on sand. Soon as the spring came I couldn't stand it because the sand fleas were so bad. And then we lived in the granary. It was twelve by twelve, and we lived right beside the new house we were fixing. Oh, the difficulties was to get ahead. No money. After we were married a few years … you couldn't sell no grain and I went to the relief

officer to get something so as I could feed the family. He told me I wasn't eligible. We had twenty dollars in the fall of the year … and got cereal and stuff like that, flour. And with what groceries we had we lived through the winter. It wasn't easy. That was the biggest difficulty. And, things was to stay alive and have enough to live on. And … people would help one another … I got fed up … I sold the grain for debts … well, I didn't sell it … traded it for debts. Leaving the farm was kinda hard. Ma was in the hospital having an operation … and that is about all I can say.[23]

Illustration 5.4 shows the poster for the auction sale at the Kerpan farm (the name is misspelled). After ten years of struggle, clearing the land of trees, producing reasonable crops for which there was too little market even in wartime, in 1942 Gabriel Kerpan took his family to British Columbia, where he learned a new trade. He and his children never returned to farming.

### *The "Female Frontier"*

The letters of Mary Percy as well as the foregoing comments of the settler wives give us a glimpse of what has been, until recent years, the largely unsung story of the "other half" of the frontier experience, what Langford has called the "female frontier."[24] The women's stories have been found in many kinds of personal documents, such as diaries, letters, and reminiscences for family members. Across the boreal margins many women experienced that "life on the frontier was made up of hard labour, a minimum of comfort and a constant battle with the elements. It was a lonely existence, with Walter out of doors early and late, not another shack close enough to send out a cheering ray of light, and neighbourly visits a rare occasion … the future looked duller and more oppressive than the life I had shed. I was bored, and homesick as well"; and Gertrude Chase added that she had seen no women for some months, and had not been off the homestead for over a year.[25] From the British Columbia corner of the Peace River country, Esme

Illustration 5.4
Auction poster for the sale of the Kerpan farm, Meath Park, SK, April, 1943; the last phase on the boreal frontier for many erstwhile farmers.
*Source:* From private collection.

Tuck pointed out that "To homestead is to be drenched with rain, caked with mud, choked with dust, chilled with cold, warmed by the sun, to rise early and go to bed late, to wonder whether roads and railway will ever come one's way, whether one has come to the right place or not, what the future holds in store for oneself and one's children, to be tired, to work, to laugh, to help the other fellow and always to hope ... At times the cold was frightful. 'Stabbed like a driven nail' is a true description ... When a wind was blowing it did not matter how many garments I put on, I was perished ... Many a time I have taken a hatchet to chop off a pound of butter to put on the breakfast table."[26]

The battle with the elements could take a number of forms in addition to the infamous drought and frost. Many pioneer environments across Canada, such as that of the Kinderwaters were plagued by grasshoppers that seemed to multiply in wild or abandoned land. Although more notorious further south, hail was also experienced in the boreal forest and at times its dismaying destruction stimulated the poetic muse: "Thirty-five acres we had, seeded to wheat, the land cleared from bush on our homestead. Dad spread his arms over the heads, and the heads nodded in the breeze ... 'It will put us on our feet,' Dad said, 'We'll pay all our bills.' A rumble from the west flung the words back to him ... The black clouds boiled, blotting out the sun. The hail rode in on green wings of destruction. There was a moment of stunned silence. Then Mom handed me a pail. 'Gather up some hailstones,' she said, 'We'll have ice cream for supper.'"[27] And Margaret Thompson, writing from half way between Edmonton and Lesser Slave Lake in 1921, reported that "When we arrived at our barnyard, we saw hailstones floating in rivulets of running water. There were a number of drowned fowl, and young pigs squealing and attempting to go for shelter. What we could see of field and garden was flattened beyond hope. Trees were stripped of leaves and the silhouettes appeared like autumn scenery."[28] Loneliness, fatigue, fear, and depression were common themes in the lives of homestead women, encapsulated by the comment of one woman far into the boreal belt in northern Saskatchewan: "I've quit wanting anything for myself, except the things to make it better for the babies."[29]

### *"When the Going Gets Tough"*

One of the most promising areas for settlement in the 1920s, with a cachet almost equivalent to that of the Peace River country, was north and east of Prince Albert, Saskatchewan, where the Kerpans located. The provincial government began surveying and opening land previously reserved for timber. A large, especially popular section for farming stretched approximately 130 km from Spruce Home straight east to Nipawin, and began to be filled rapidly at the end of the 1920s, partly by "dried-out" farmers from the prairie. The railways and the Soldier Settlement Board were active in encouraging the entry of settlers. Despite the promise of the area, it was not free of abandonments. As the evidence to the commission, below, indicates, the popular saying "When the going gets tough, the tough get going" may be appropriate to some of these stories.

Julius Androchowiez served as an agent for the CNR (as mentioned in chapter 4), and testified before the Saskatchewan Royal Commission on Immigration and Settlement (a response to the intensifying crisis of dryland and other marginal farm failures). Appearing in Prince Albert on 25 April 1930, Androchowiez reported that:

> Since 1927 I have settled 34 families on purchased land north of this City ... the first settler who came in under the Three Thousand [Families] Scheme ... had a capital of $350.00. I sold that man a quarter section of land a mile and a half north of Henribourg for $1600.00. He was obliged to pay $300.00 down ... and was left with $50.00 and three children. He had no furniture and nothing to buy it with. $50.00 was all he had to buy stock with. He is a Pole, and that particular man can stand inspection today ... He learned English, his wife speaks English, and his children attend school. He told me yesterday if anyone offered him $3200.00 cash he would think twice before selling it. I sold that to him in 1927 ... I would say that 80% of those I settled on purchased land are making good, 20% will very shortly need my attention to be replaced on cheaper quarters as they cannot make ends meet.[30]

Another knowledgeable observer from fifty kilometres further east in this late frontier had a more critical view of the agricultural promise of the district (cited in the previous chapter). He told the Royal Commission, meeting in Nipawin on 19 February 1930, that he had homesteaded ten years earlier and had been followed by five to seven hundred other settlers. In 1925:

I was hired by Mr Bellamy [of the Soldier Settlement Board] to go with him. We covered the country very thoroughly. We went from Nipawin as far north as Township fifty-four, East to Range twelve, West to Range twenty-two. [Commissioners Reusch and Neff asked questions, indicated by Q, followed by the witness's answer.]
Q: Did you find considerable land unsuitable? Practically everything the Government survey said was unsuitable we found [to be] correct.
Q: Has there been any settlement on this unsuitable land? Practically all the settlement in the past two years has been on land not suitable for agricultural purposes.
Q: What would you say the average cost of clearing land in this area is? The average cost … can be place[d] at forty dollars an acre … Some will cost more. One man can clear three or four acres a year if so minded … Some have been in eleven years and only have two acres cleared. I have been there ten years and have hired a lot of help and cleared seventy acres. If I had had to do it alone, I would have about thirty acres cleared … They should not be allowed to homestead in this country without capital, although some made a success of it … The minimum amount required would be eight hundred dollars. To relieve the unemployment situation the Government should classify all these lands into lands suitable for homesteading and that which is not, and then slice down about one hundred acres on each quarter [section] and charge it up against the land …
Q: What proportion of the homesteaders you know are successful? … In our school district there are eighty-seven quarters. Eleven of the original settlers are still there.
Q: Do you mean that the other fellows threw up the sponge? I mean that they threw up the sponge and I think I would be safe in saying that those conditions prevail all over.

Q: How does the standard of living compare of the homesteader here and the homesteader on the prairies? I do not know much about the prairies. The standard of living of all nationalities here [in the boreal forest] is about the same. They are just existing.
Q: What is your idea as to the question of homesteads? In my opinion it is a poor policy ... Take the man starting on a bush farm. He builds a little shack then he has to get out to make a living for his family. He works all winter hard. You know what wages are in the North Country ... In Spring he plants a garden and goes out again. He puts in four or five years and has one or two acres broken. He never gets ahead.
Q: Of the homesteads taken in 1929, how many, do you think, will stick? I figure about fifty per cent. will stick.
Q: Why? Because they are a different nationality and they are forming a colony. They are mostly Ukrainians and they will be satisfied with a lower standard of living. They are coming from practically intolerable conditions in their own country. The standard here will be a little better.
Q: In the area of land you covered in the survey how much was fit for homesteading? In an area of roughly one hundred thirty-two miles by twenty-four miles, I doubt if over thirty per cent was fit. Of the other seventy per cent., fifteen per cent. was merchantable timber and the rest was small jackpine and spruce not suitable for anything.
Q: In your opinion to what extent has the homesteader increased the fire risk to timber limits? Well, there is only one way to clear this land and that is fire. If you are going to make a farm you don't care for timber.
Q: To what extent do fires get away from homesteaders? The moment he lights a fire it usually gets away from him.[31]

Throughout the boreal fringe the excessive variability of surface materials and vegetation, the severity of climate, the isolation, and the homesteaders' coping strategies were repeated from one pocket of incursions to another. This produced a late frontier that was everywhere daunting, with broad similarities in approaches and techniques, but in the fine detail showing evidence of unique individual efforts.

It was a new territory entered with enthusiasm, but a careful observer could see that most of it offered only an uncertain promise of a brighter future.

### *Testing Margins* en bloc: *Group Ventures*

Group settlement was thought, by the 1920s, to offer the best hope for farming success in the newly opened land. As it was phrased in a study of 1926, "The modern tendency in colonization is toward group settlement." C.A. Dawson claimed in one of the classic Canadian Frontiers of Settlement volumes in the 1930s, "It seems clear that group settlement as contrasted with individual settlement makes for greater residential stability ... partly because of joint effort which emerged quite readily in these homogeneous groups."[32] Even homogeneity, however, had a variety of expressions, as Dawson recognized. At one extreme were the groups that fervently sought isolation, even to the extent of relocating in search of it, then those that cooperated with the surrounding society and its facilities, and at the other extreme were those that gradually dissolved into the mainstream and lost most distinguishing characteristics in the process. The boreal forest frontier was not generally chosen by the well-known, strongly exclusionary sects prior to the 1940s, although the Doukhobors struggled for a decade in the bush country of northeastern Saskatchewan. By the time this study begins, the Doukhobor experiment was well along the road to disintegration. There was also the small, ill-fated Mennonite settlement in the western sector of the Great Clay Belt in Ontario; and the *Regina Leader-Post* and various other newspapers were tracking a group of Mennonites relocating northwest of North Battleford, Saskatchewan, in 1934.[33]

The typical group that found its way to the boreal margins in the 1910s to 1930s was brought together by an organizing agency, or by some common affiliation, generally religious or ethnic. As years passed and difficulties mounted it was common for the adhesion to the original aims to weaken. This was even true of the longstanding system of colonization, combining government support and zealous church activity in establishing parishes and a social structure, that had been developed in Québec. As Courville found in present-day Abitibi, the settlement founded on an agricultural dream nourished by church and state, while lasting notably longer than most other planned settlements in the boreal forest, has diminished to "no more than vestiges clinging to the best land."[34] Although northern settlement in Québec was relatively more successful than elsewhere in Canada, over time many people reluctantly abandoned their boreal farms and in so doing enriched a reservoir of arcadian nostalgia that has found outlets in literature and film, such as "Kamouraska," "Trente Arpents" ("Thirty Acres"), and "Les Filles de Caleb" ("Caleb's Daughters").

The *blocs* analysed by Dawson were the most prominent in the western farming expansion in Canada. They accounted for hundreds of individuals each and, significantly, flourished prior to the late frontier of interest to this book. Dawson's volume on group settlement dealt with the Doukhobor, Mennonite, Mormon, German Catholic, and French Canadian cases in western Canada. The groups found on the late northern frontier were much smaller and in almost all instances less tightly knit. For purposes of this discussion, an eligible group would be larger than an extended family, let us say at least fifty; but few would have been larger than a couple of hundred souls. Some examples follow.

### *Soldier Settlements*

Some early pushes into the boreal margin were a result of government attempts to accommodate servicemen demobilized from World War I. The first legislation for this purpose was passed in 1917, and with its successors formed a body of federal and provincial "soldier settlement" acts. Similar legislation was also fashioned in Australia and New Zealand and, as in Canada, came to include British as well as indigenous military personnel. This legislation was designed as a reward to the servicemen for their huge sacrifices on behalf of their country. In many cases the reward became a burden: post-war inflation and the varied competencies of the administrators of the legislation could lead to serious difficulties, and to abandonments, for servicemen in all the corners of the Empire where soldier settlement was introduced. On their parts, the servicemen, heroes though they were, also proved to be ordinary people with ordinary human frailties, though many were dealing with more than ordinary recent experiences: "some of them had suffered severe wounds during the First World War, and all, although not immediately evident to the human eye, were suffering the debilitating effects of the long and strength sapping duties in and around the trenches under the most atrocious conditions imaginable."[35]

The resulting drama of a returned servicemen's settlement was played out in one documented Ontario case named the Kapuskasing Colony. It was located in the northwest section of the Great Clay Belt, 120 km west of the town of Cochrane. Typical of the boreal margin growing degree day range, Kapuskasing records 1339 growing degree days.[36] The legislation that allowed for the establishment of an organized settlement was passed in the Ontario House of Assembly in April 1917; three years later a commission of inquiry was set up to investigate what had become a broadly troubled situation in the colony. There seems to be a touch of sarcasm in the report of the commission of inquiry, in its list of the attractions of the area for the former government minister who championed the proposal in 1917: "(a) The quality of the soil … a deep clay, (b) The beauty of the landscape, (c) The … agreement with the Dominion Government … for … an experimental farm … opposite the Colony site, (d) The desire to establish a Colony…of a well-disposed group of returned men, (e) The water powers … available for industrial development."[37] These attractions were to turn sour during the few years the colony survived, except for the experimental farm and the perennial water power that were independent of the colony. Of course another thing that did not change was the soil. As figure 5.2 shows, the term Great Clay Belt is a gross oversimplification as far as the surface conditions are concerned: instead of a broad expanse of similar clay soil that might be useful for agriculture, it is an astonishing mixture of soils, gravels, and wet mosses, with scattered enclaves of tillable land. Good farmland (classes 1 and 2) is lacking, and the patches of classes 3 and 4 land, which might produce a hardy crop, are compromised by a complex surface or wetness or a severe climate. The large incursions of organic soil would interfere with farming activities. Similar patterns continue eastward to characterize most of the Cochrane District and, as figure 5.1 illustrates, the Abitibi portion of Québec. Generally it was not practical to undertake farming more than 12 km from the railway that crossed the Great Clay Belt from east to west.[38] Obviously the climate, with its cool summers and cloudy autumns, does not make up for the limitations of the land.

The Kapuskasing story provides an object lesson on the difficulties involved in establishing a group farm settlement. One of the recognized experts on urban and rural planning, Thomas Adams, described Ontario as "probably the most advanced" of the provinces in preparing for the return of servicemen to the land, specifically to the clay belts of northern Ontario.[39] Ontario passed its Returned Soldiers' and Sailors' Land Settlement Act four months before the federal government passed its comparable Soldier Settlement Act in August,[40] and began preparations for the Kapuskasing scheme about the same time. It was visualized that there might eventually be some hundreds of residents, but the growth would be regulated. The planners had the advantage of a recent upsurge of scholarly interest in how best to prepare, including practical examples from British experience and the nearby guidance and encouragement of the Commission of Conservation in particular. The returned servicemen who applied were to be selected for their supposed relevant skills and experience, and to be sent in a party of twenty to thirty men for a short period of instruction in northern Ontario. After instruction the party would go to the settlement area to begin co-operatively clearing ten acres on each of the allocated lots and, at the same time to build up interrelationships that would form the matrix of a community. Other similar work parties would follow at intervals. Each settler was allotted an eighty-acre block (later increased to one hundred acres), and was given assistance in acquiring domestic livestock and equipment, as well as a $150 grant toward a house. A loan up to $500 was available. The land being offered was all close to the National Transcontinental Railway, east of the Kapuskasing River, and near the experimental farm being developed by the federal government.[41] The first party of twenty-four men was sent north in June 1917. The commission of inquiry learned that from an early date "the best laid plans" had begun to go awry:

Instruction at Montieth [Experimental Farm] did not progress as had been expected. The men were anxious to go forward to the settlement at Kapuskasing, and resented being held at Montieth doing work that they thought could be quite as well learned at their new home. In some cases the prospective settler did not seem to value the opportunity for instruction nor to respect the authority of the instructors. As a result, toward the latter part of July, the group left Montieth for Kapuskasing … The original plan was for the men to work in groups under competent instructors, slashing, burning, clearing, and building, but this did not long prevail, as the more ambitious and energetic of the settlers did not view with favour the work done by their less able or less willing companions … The experience of 1918 showed that, while the skilful and energetic made progress and money, others were not getting their holdings cleared satisfactorily, and were … remaining "on the scheme" unreasonably long … as there is not now sufficient remunerative work to be had in the Colony for the settlers, it is not surprising that resentment is felt by those whose ten-acre clearings are only about one-fourth the size required to produce a living for a family by agriculture alone … the Government announced in the press that it had granted a concession for pulp operations over a large area adjacent to the Colony site, and it is to be regretted that for financial and other reasons the work has not been carried to completion. If the mill were in operation to-day a good deal of the distress in the Colony would be relieved.

To understand the difficulties of the Colony to-day, it is necessary to bear in mind the nature of the forest and soil at Kapuskasing and in the clay belt generally. About seventy per cent. of the land is covered with a forest growth of spruce, 2,500, or more, trees per acre ... the largest trees are about twelve inches in diameter, but most of them are much smaller. On the other thirty per cent. there is larger spruce, poplar, and balm of Gilead [balsam poplar] ... To clear such land economically, the marketable timber is removed, the remainder slashed, usually into neat windrows, and, when thoroughly dry, burnt. If what is called a "good burn" is secured, most of the poles and moss will be gone. The removal of the green stumps is a very costly operation ... Those who have had considerable experience in clearing this land advise sowing grass seed on it as soon as the brush and poles are burnt, and leaving it in pasture until the roots are so rotten that the stumps can be easily gathered and burnt. This also gives an opportunity for the water to drain away, so that the soil is drier and warmer ... this takes time, four or five years ... During this time of waiting, the settler, if without capital, must earn most of his living by other means than agriculture ... clearing in a large way is absolutely necessary if tender crops are to be grown, for the reason that the moss-covered soil retains the cold and frost of winter far into the summer months, with the result that summer and early fall frosts often strike small clearings ...

the ordinary pioneer, as soon as he gets a shack built, some little clearing done, a few potatoes and vegetables planted, looks about for work that will enable him to keep his family in the meantime and over the following summer. He usually finds this in a lumber camp – perhaps many miles from his home ... On the other hand most – if not all – of the Kapuskasing settlers tacitly inferred from the pamphlets published by the Ontario Government, and from conversation with its officials, that the responsibility for providing them with work and good wages at their own homes was entirely assumed by the Government. It is to be regretted that in the various publications setting forth the advantages of the Colony to the returned men, greater emphasis was not laid on the conditions, difficulties and problems of the district. The most serious personal difficulty was the fact that many of the men and women were by their natural disposition and previous occupation and experience quite unsuited to pioneer life ... it would have been better if the fitness of the men for pioneer work had been more searchingly considered before sending them to the Colony, and even afterwards to have returned, as unsuitable, those who did not measure up to requirements ...

to fill adequately the position of ... Superintendent of such a Colony, under the conditions obtaining in Kapuskasing, a very unusual combination of qualities is required such as, a thorough knowledge of –

(a) The best [m]ethods ... of clearing that class of land.

(b) ... selling, cutting, hauling etc., of all kinds of saleable timber found there.

(c) ... preparing a good seed bed in such land.

(d) The most desirable crops to be planted at first on such soil ...

He should also have a good faculty of conveying this knowledge to settlers, he should have the ability and tact necessary to stimulate those under him to do their best, and ... have an unfailing fund of sympathy ... Your Commission has no hesitation in saying that not one of those placed in charge possessed all these qualifications, and some of them possessed scarcely more than one or two... the powers of the Superintendent were not clearly defined. Several ... cases of difficulty arising from divided responsibility came before the Commission. Conference and co-operation would have created good feeling, whereas arbitrary rulings provoked discord.

If building and equipment could have guaranteed success, the Colony should have prospered ... the Province has spent a large amount of money for very small returns, and your Commission is of the opinion that the unsatisfactory conditions should have been dealt with at a much earlier date ... not only in the interests of the Kapuskasing settlers, but also in the interest of the Province as well.[42]

The major items of cost to the Province, amounting to nearly one million dollars, had been the establishment of the

Monteith Experimental Farm, the building of the Kapuskasing colony structures, the provision of horses and equipment, the construction of roads and bridges, and the clearing of land. The general recommendation of the commission was that the colony should be put on a strict business-like basis; that unsuccessful settlers should be assisted to transfer to a different occupation in another part of the province; that settlers choosing to remain should be encouraged to achieve early independence; and that the paternalistic role of the government should be ended, except for provision of an adviser and a teacher for a couple of years. Perhaps some sympathy is due the government that originated the Kapuskasing experiment because, if agriculture was to succeed in northern Ontario, the location chosen seemed to provide much of what experts considered essential. For one thing it was directly on a major railway, offering connections to markets and modern amenities, and it was next door to an established experimental farm that had been testing various agricultural regimens. Unfortunately what were not well tested were the two agricultural bases – climate and soils, about which the scientist Fernow had communicated his concerns in his report to the Commission of Conservation six years earlier.[43] In both 1918 and 1919 the crops on the settlers' new plots had been wiped out by frost. Kirkconnell's review of the colony claims that 101 settlers had been put on the land but that in the winding down of the colony in 1920 only 20 elected to remain. Murchie records that twelve years later eleven of the original 103 were still in the colony, and Arthur Lower, looking back in 1935, concludes that the attempt to settle returned soldiers there "proved a complete failure."[44]

Another soldier settlement scheme was taking shape about this same time in a newly opened part of Saskatchewan. Because Crown land in the Prairie Provinces was under the jurisdiction of the federal government until 1930, this scheme, known as the Porcupine Plain soldier settlement, was planned in Ottawa and managed by the Soldier Settlement Board. The area was carved out of former forest reserves 200 km east and south of Prince Albert, stretching toward the higher land that straddled the eastern boundary of the province and, despite a sophisticated approach to some aspects of the planning, the invariable challenges turned out to be widespread poor drainage and colder temperatures with a markedly shorter frost-free season related to the general rise in elevation. The soils that provided some hope for agriculture in the area are classes 2 and 3 north of Porcupine Plain village (see the Canada Land Inventory map of classification of soils for agriculture, figure 5.3; Porcupine Plain is on the southern border and Prairie River is at the top right). The hatch marks indicate that in large sections the surface is a complex of slopes, and land with wetness as a limitation amounts to about 20 – 30 per cent of the area, without including the organic classification. The C subscript represents the ubiquitous climate limitation. Although a new railway line skirted north of the settlement, providing an off-loading point in the woods at the oddly named Prairie River, there was no all-season road to get the settlers the ten to twenty or more miles to the railway even six years after the opening of the project: "Notwithstanding the huge sums that have been spent previously on this road, there are still approximately 7 or 8 miles that is nothing more or less than a quagmire."[45] The investment of tens of thousands of dollars by the Soldier Settlement Board into work on roads and bridges suggests that the area had more than the normal range of difficulties for pioneering. There were a number of worrying similarities to the Kapuskasing Colony, such as the heavy bush requiring clearing of sizable trees, the complex surface conditions, and the wet land. In addition, it was far distant from "home office" decision-making and sources of aid. The Porcupine Plain soldier settlers became exasperated with the administration of the scheme, and in 1925 held an angry meeting in somewhat the same way that the Kapuskasing veterans confronted the Ontario legislators. At the end of 1920, 101 soldiers were settled at Porcupine Plain; through alternating abandonment and

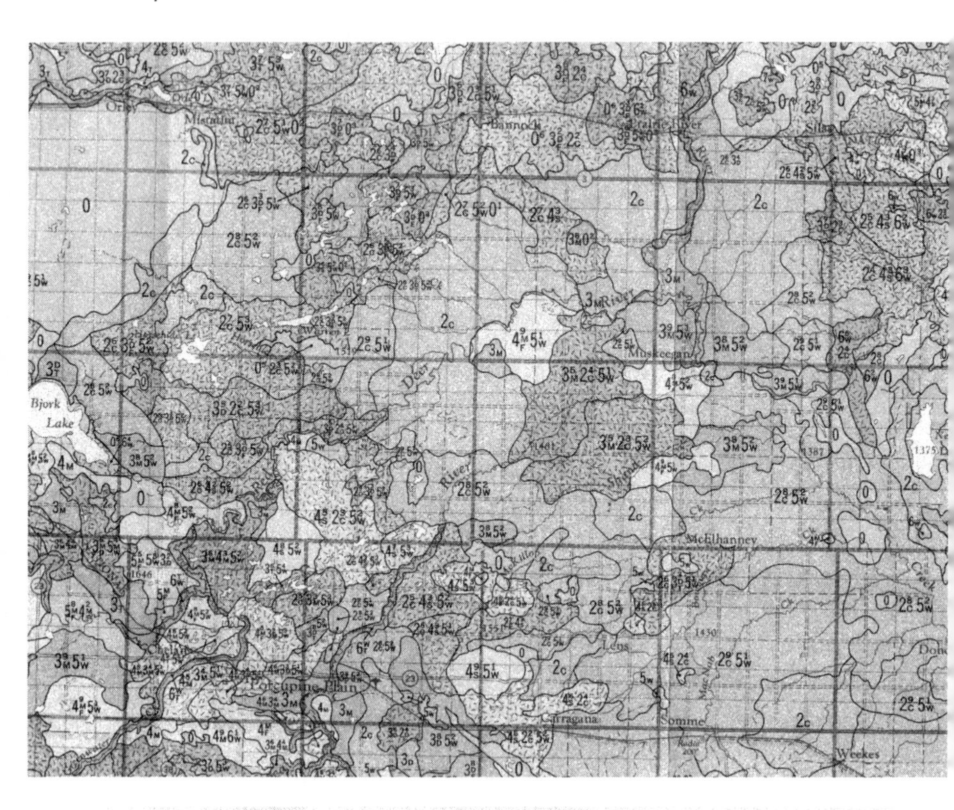


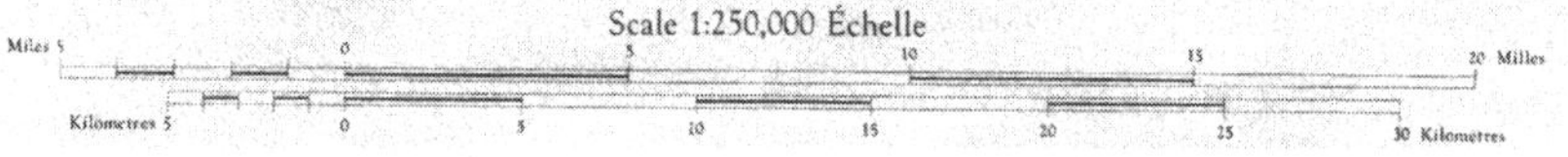


Figure 5.3
CLI assessment of soil quality for agriculture, Porcupine Plain, Saskatchewan. Although there are a couple of patches of relatively homogeneous land, this area, with the boreal margin in general, shows the arresting variability in land quality that immediately suggests difficulties. An additional concern is the subscript letter C that warns of climatic hazards for agriculture. Also see appendix B.
*Source:* Part of CLI map 63D "Hudson Bay."

new entries, there were 150 at the end of 1923 and 139 in 1925.[46]

In July 1919 the Porcupine Plain project was opened, with each applicant soldier acquiring his lot through a draw at Prairie River (initially intended to be a "model" town) on the railway line. One of the veterans in the original draw provides recollections, forty-six years later, of the foundation of the project:

... on opening day, a large, new, revolving churn was used and all names on entry slips were placed in this and then drawn, one by one ... Then the trek to the Settlement began. The first stage was seven miles to the Red Deer River which had no bridge and a poor crossing ... When the settlers reached the river, then the first casualties commenced when they saw the water flowing swiftly by the crude crossing, between four and five feet deep and very cold as well. It meant stripping and wading across, holding our clothes over our heads and several decided (perhaps wisely) that this wasn't so good and quit right there ... That night we sat around a large fire (it froze nearly every night) which helped to keep the mosquitoes down, told each other what we planned and then out came a violin ...

there was a temporary pause in settlement until the spring of 1920, when ... we all came flocking in and the settlement was away. Everyone had to get a building up of sorts, which was mainly of logs and remained that way for several years ... In 1921 the log walls were erected for a building intended for a hall, but with families increasing and completely isolated at the time ... this building was changed into a hospital and was taken over as a Red Cross Outpost ... In 1921 a rural mail route was started from Prairie River to four points in the Settlement [and] the question of education was solved swiftly and neatly by two school boards ... They decided that [one] would build the school and both districts join in the operating costs ...

In 1924 St. Andrew's log church was built ... Things were getting better, but roads were woefully few and bad in general, but when conditions were possible, a little work was being [done] on them which was only a small amount in the fall, if it was dry enough ... In 1926 ... it was a tremendous lift to the settlement to have a through road to Prairie River at last ... Our community hall was started in 1923 and completed in 1924. It was made of logs of course, but was large with plenty of room for all occasions. Around 1925 we had formed our branch of the Canadian Legion, the fourth formed in the province ...

> In 1927, British settlers were brought out, several of whom settled in this district on farms taken over by the Soldier Settlement Board from settlers who had quit their farms … the Depression hit everyone in general and us in particular. We had fought our way up after ten years to a point where we had something to sell and now you could hardly give it away.[47]

The soldier settlements at the end of World War I were groups only in the sense of having a superimposed integration of the projects. Even if these group experiments survived, it was only a matter of time – sometimes very short – before wide-based collaboration faded away. Differing skills, strength, interests, and individual land ownership led to settlers going their own ways, and group cohesion remained only as a memory of "the old days."

There were successful settlers attributable to the various Soldier Settlement Acts, some of them in the Porcupine Plain area, but it is hard to see this approach as a good use of the large sums invested. In a survey of the soldier settlement in the British Empire between the world wars, Fedorowich concludes that everywhere it had serious defects. In Canada, of the 25,000 soldier settlers who took loans from the Soldier Settlement Board after 1917, only 43 per cent were still on their farms in 1931, but 36 per cent of those had *no equity* in their farms.[48] Manitoba was on the same general trajectory in 1926, eight years after passage of the Act, when Murchie and Grant (as noted above) found that over 43 per cent of the soldier settlers in that province had given up their farms. Murchie concluded in 1933 that the Canadian soldier settlement schemes showed only meagre success.[49] And when, as Thomas Adams noted in 1917, "only a very small percentage of the twelve thousand returned men so far … was willing to go on the land … It seems to be assumed that a larger proportion of the soldiers will go back to the land than have come from the land … Out of 346 soldiers who had returned to Alberta only six had signified a willingness to take up farming, although a number of the returned men had been farmers before they enlisted,"[50] should we be surprised at the "meagre success"? The soldier settlement schemes were a prime example of governments pushing misguided and commonly unprepared people into unpromising locations as a means to resolve the returned servicemen "problem." The blame should at least be shared, as Powell would have it, because "the so-called 'failure' may be a misnomer except in so far as the administrative and political system disappointed and distressed the community in general and the soldier settlers in particular."[51]

### *The Blacks of Amber Valley*

One group on the late frontier had cohesion that was brought about by racial distinctiveness strengthened by experiences of persecution. The African-Americans of Amber Valley, in northcentral Alberta, came into Canada at the beginning of the period covered by this book. They had suffered persecution in the rapidly developing southern plains states in the United States and had seen Canada (perhaps in the afterglow of the so-called Underground Railroad) as a place of promise. They came into western Canada in a number of small contingents, mainly around 1910. They discovered that prejudice had not been left behind, and more entries by "Negroes" were virtually cut off as a result of newspaper agitation and petitions in the West, followed by an immigration act of the Liberal government.[52] The Immigration Act of 1910 allowed for barring from the country "immigrants belonging to any race deemed unsuited to the climate or requirements of Canada"; and under this section 38, shortly before its electoral defeat in September 1911, the federal cabinet was considering an order-in-council to ensure that "the landing in Canada shall be ... prohibited of any immigrants belonging to the Negro race, which race is deemed unsuitable."[53] A significant proportion of the American blacks from Oklahoma chose to continue a rural way of life and settled in bush country twenty miles (32 km) east of Athabasca, Alberta. In this area, called Amber Valley, some dozens of black settlers took up land in the early 1910s and, through a typical pioneering progression, the group increased in number to reach a high point close to four hundred individuals in the 1930s. By that time the community was well established, with schools, churches, and businesses; one member had been appointed to take the 1921 census in Athabasca district, perhaps a sign of rising status (though some overt discrimination emerged).[54]

The success recognized in hindsight was gained through probably more than the typical allotment of pain, because these were southern and prairie people with next to no experience of bitter cold or of chopping down trees (see the land conditions illustrated in figure 5.4). The outstanding characteristic of the Canada Land Inventory map is the dominance of organic (generally boggy) soils. The strip of 3 and 4 class soils around Amber Valley is almost enclosed by the non-agricultural organic areas and Flat Lake. Typical of the boreal forest case studies is the general negative influence of the climate. The difficulties of clearing land and surviving the winters loom large in the reminiscences:

We had to build our houses out of logs … We hewed the logs and put them up and then plastered them with clay, but the clay would crack and the cold in the winter and the flies in the summer would come through the cracks. Now the Ukrainians were used to the cold and knew how to build good houses, but we didn't … Sometimes the colored folks would hire the Ukrainians to help with their homes … The roof was made out of poles with sod on top. The sod was good insulation, but wasn't very good protection against the rain. We used to say that it rained outside one day and inside four … there were still lots of holes [that let insects in] … the sand flies were the worst. They would crawl in your ears and nose, and around your eyes, and bite and torment you to death. Most of the time we had to wear mosquito netting reaching from our hats to our waists. The sand flies were terrible on the animals, too. We had a mixture of axle grease and other things that we would smear in their ears, and around their

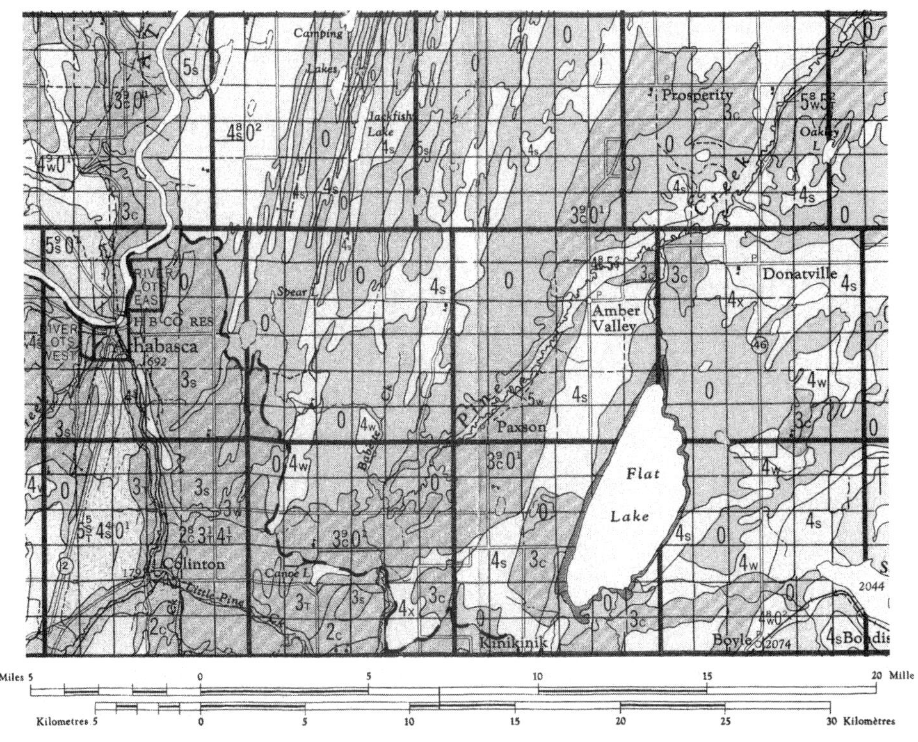


Figure 5.4
CLI assessment of soil quality for agriculture, Amber Valley, Alberta. Some class 3 land appears, but is surrounded by a large expanse of organic soils.
*Source:* Part of CLI map 83I "Tawatinaw."

eyes and nose, and on their flanks, so they could stand to work. The mosquitoes were bad too, but they didn't stay at us so constantly as those sand flies.[55]

Clearing enough trees to put in a crop was so difficult and slow that for families to survive in the early years the men had to take off-farm work, especially through the winter. Sometimes this work was at a considerable distance, and over the years it could cover a wide range of occupations such as digging for building foundations, shovelling coal at a depot, or working on railway construction or in a logging camp. In one of the first winters J.D. [Jeff] Edwards worked for a logging company, six ten-hour days a week:

I came home every week. After work Saturday night, I would walk the nine miles to Athabasca, load some groceries on my back, and then the twenty miles to home. Sometimes it was pretty cold – forty below or colder. I would chop wood on Sunday so that my wife could stay home and milk the cow and take care of the baby. Then I would walk back in time for work Monday … I wore moccasins with rubbers over them, and sometimes felt boots.[56]

A number of the Amber Valley men worked winters freighting goods. Jeff Edwards recalls:

it would get real cold when you were travelling, you couldn't sit on your sleigh, you'd freeze. Anyway, when the driver walked, it lightened the load for the horses. Our teams would learn to follow each other. You never freighted alone, it wasn't safe … If you had a good team you could get off your sleigh and three or four of you would walk together and visit. [Another freighter adds:] There were stopping places along the way that consisted of an old log shack and some barns for the horses … It cost me a dollar a night for hay for my team, and I carried my own oats. They were quite expensive … There wasn't much farm land cleared yet and so oats were expensive … when I got enough of my land cleared to make a living, I gave up freighting.[57]

As with most frontier settlements (Dr Mary Percy's territory being an exception) no doctor was available to the Amber Valley people for years. Traditional remedies and mutual help were applied in cases of sickness, but there were losses to the 1918 influenza epidemic. Assistance in childbirth was provided by a local midwife: "Mrs Broadie was a fine midwife. She would come to the house two or three days early, and stay a couple of days after the baby was born. Then she would move on to some other house where she was needed. She took care of ten of our eleven children."[58]

The Amber Valley blacks were in a way circumscribed by a somewhat hostile social environment, as they were by a hostile natural environment of swamps, but they were not internally homogeneous to the same extent as the religion-based communal groups such as the Hutterites or Doukhobors. Amber Valley people were free to go their own way, different church denominations were brought to the settlement, and men went in different directions in search of outside income in varied occupations. At times they must have felt strongly, perhaps even more than contemporary Québécois pioneers half a continent to the east, that "the unfeeling and imperious land was the lordly suzerain whose serfs they were, paying their dues to the inclement weather in the form of ruined harvests, subjected to the forced labour of digging ditches and clearing away the forests."[59]

By the time of World War II many of the young people were leaving to go to jobs or military service, and the founding generation members were beginning to retire. The community had worked through what must often have seemed insurmountable obstacles to a level of success that allowed them to begin to disperse.

### *Ethnicity and Religion Together*

The cohesive power of an exclusionary religious affiliation characterized some of the largest and most successful of the group settlements in western Canada around the turn of the twentieth century. These *bloc* settlements, featured in some of the classic works of the 1920s and 30s on the Canadian frontier, were able to find good land in the grassland or parkland belts, and did not enter the northern forests. This applies even to the late-arriving Hutterites. But some ethno-religious groups entered the boreal margin during the period covered by this study. The Edenbridge Jewish colony originated in 1906 and expanded in the following decade, the Reesor Mennonite settlement began forming in northern Ontario in 1926, and the La Crete Mennonite colony was started just before World War II.

To describe the Edenbridge settlement as "religious" is perhaps stretching that term too far. Edenbridge was like most of the Jewish settlements in western Canada, more economic than religious in intent.[60] As Waddington shows, a stronger passion in Edenbridge arose from the socialism of the eastern European settlers, such as Michael Usishkin, whose memoirs she discovered: "there was a strong minority of secular Jews devoted to Yiddish language and culture who had been politically active in the socialist movements in Russia."[61] Although settlement had begun in 1906, Waddington's correspondent, Usishkin, and his socialist friends had joined in 1910 and 11, arriving from England. At this time the "colony" consisted of three clusters of quarter-section farms strung out along 18 miles (29 km) of a north-south road, with a synagogue and community hall near the middle.[62] Waddington paraphrases Usishkin's lively memoirs:

In common with all homesteaders, the Edenbridgers built tiny one-room log cabins with sod roofs and tables and benches for beds. Women gave birth without benefit of doctor or hospital and the young husbands hired themselves out to distant farmers every summer so they could buy the barest necessities. Usishkin tells of wives and children who remained at home, gathering dead branches and breaking them up for firewood, and how, when the young mothers had to round up the cows at night, they had first to put out the stove, empty the water buckets, hide the matches and leave their little children alone in locked houses. He describes the bitter winters when newly born calves and chickens had to be taken into the one-room house lest they freeze …

Political discussion was important in Edenbridge, and there were countless meetings … They also started a credit union, a library, a Yiddish secular school, and built a hall where they staged plays and educational evenings. "We weren't yet Americanized and knew nothing of the American eleventh commandment, 'Hurry up' nor the twelfth, 'time is money,' nor yet of the thirteenth, 'business before pleasure' …"

"My first walk to the post office" … will convey something of the flavour of Usishkin's story: "The post office was about the size of an average bedroom in an ordinary house ... Small as it was, this single room was partitioned off into three smaller ones. One was the so-called 'big room' where the public gathered to wait for the mail. In the centre of it [stood] a black iron stove that drew everyone around it like iron filings around a magnet. There was no furniture, but none was necessary. People sat on sawed-off logs, and if it wasn't too cold, they spread themselves out on the floor like so many geese. One seldom met women at the post office ... since they couldn't easily make their way through the swamps ... the whole building was so flimsy that it could have been blown over by a breath of wind. But in those days the bush surrounding the post office was very dense, so that even a paper house would have survived. In the waiting room I encountered a number of Jews whom at first I couldn't fit into the context of any European country. There was a mishmash of faces from all over Europe, Africa and America ... there were the overalls, indigenous to Edenbridge itself. They consisted of a veritable chessboard of patches. And every man without exception had a beard."[63]

Robert England, in the mid-1930s, referred to the colony as Lithuanian, some members of which had been in the Transvaal. Although the group was decreasing, it still numbered over a hundred residents and maintained a dramatic society and news sheet. Katz and Lehr record that after forty-five years, in 1951, thirty Jewish families still lived around Edenbridge, whereas most of the fifteen Jewish colonies across the Prairie Provinces were in the process of abandonment or had already disappeared.[64]

In northern Ontario, in 1926, five or six years after the Kapuskasing colony of ex-servicemen had fallen on hard times, a group of Mennonites chose to settle even farther west in the Great Clay Belt. The group was a small part of the large exodus of Mennonites that left Russia after the revolution. The few dozen who homesteaded took up land primarily in Eilber Township (see the population chronology in figure 2.4), and their settlement became known as Reesor. One of the leaders of the community sent a letter to Premier Ferguson in 1926 expressing their "hope to carry out here, in Northern Ontario, a future with a religious life such as we formerly enjoyed in Russia … The settlement numbers about seventy souls."[65] The informal history of the settlement traces an almost classic boreal margin story from the excitement and travail of the early days through years of effort and disappointment to the discouraging loss of neighbours and relatives, and finally dissolution. By the end of the 1940s, after fewer than twenty-five years, the settlement was well on the way to closing down.

The Reesor colony, in practical terms, began "at 11 p.m., on a pitch-dark night, in a heavy rain, greeted by scurrying rabbits and swarms of hungry mosquitoes [when] the group of nine settlers got off the express which had been stopped just for them in the midst of nowhere, to begin their self-appointed task of hewing a new home from the primeval forest."[66] The next day the land agent arrived from Hearst, about 45 km to the west, and

> One by one the men selected homesteads … and got them registered in their names. When this was done, they began building on their own lands, usually working in pairs, since one man could not handle the heavy logs alone … When we sat beside our makeshift tables and partook of the meals we had cooked over the open fire, rabbits and squirrels would come right into our midst to snatch food remnants that were thrown to them. They had no fear, for they had not yet come to know man … One day, after the main house walls stood ten feet high and we were busy on the gable and roof section, Dad slipped on a beam and fell, hitting the upper rim of the wall with his chest … It was impossible to call a doctor, for the nearest one was 30 miles away and there were no roads … But a miracle happened! Dad spent a few days on his sickbed in the shanty, then got up and continued to work on the house. Later it was discovered that he had fractured two ribs … When the houses were finally finished after three months of arduous labour, preparations were made to have the families come up from southern Ontario … the coming of additional families from the western provinces and southern Ontario

... ushered in the beginnings of community life ... One of their first concerns was the establishment of a school section."[67]

Their school, with a Russian-born Mennonite teacher competent in German and English, was registered as S.S. No.3, Eilber. With the coming of winter, the men engaged in cutting trees for paper pulp production. The history records that all the lots in the district were thickly covered with trees, but only half the area provided the large black spruce valued as pulpwood. Once the pulpwood was cut off the lots, and livelihood depended on agriculture, commitment to the settlement began to falter. Families began to leave, until little more than half a dozen adherents remained. "It was agreed among these few to continue Sunday services in the homes of the two families on a bi-weekly basis. This was kept up with decreasing regularity until they were discontinued after November 13th, 1949."[68]

The Old Colony Mennonite settlement of La Crete, in far northern Alberta, was at a late stage – but not the end – of a chain of moves by members of this conservative group. Beginning in 1936, the settlement took shape in an area of aspen poplar, with patches of open bush, and level land with apparent agricultural potential. In this environment, clearing the trees and burning the debris, then waiting till ploughing could make the field ready for crops, was perhaps easier than in the heavy forest parts of the boreal margin, but could still take three or four years. A major desirable feature for the founders of the settlement was the almost complete separation from the outside world, the only connection being by a seasonal boat on the Peace River.[69] The early Mennonite settlers came from Saskatchewan and Mexico and, having had generations of pioneering experience, they quickly created a community structure. By the late 1950s there were 1,500 members. Disruptions to the Old Colony way of life gradually invaded, including a road, a public school, and less conservative Mennonites, but the settlement flourishes today despite significant exoduses over the past forty years.

## *Group Settlement Makes a Landscape: A Bird's-eye View*

Through Land Office records it is often possible to reconstruct the territory over which a group settlement spread its ownership. The original homestead entries of the late nineteenth and twentieth centuries are normally preserved in provincial surveyor-generals' departments or provincial archives, and can be extracted for mapping. This kind of source has been widely accessed by historical geographers to give a unique view of group settling on the land. The square surveys of the Prairie Provinces and northern Ontario, and the rectangular surveys of southern Ontario and northern Québec lend themselves especially well to this kind of reconstruction and to the engrossing interplay of structured (property lines) versus domesticated (used, "customized") space.[70]

The Northfield settlement of Norwegians is a boreal margin example of settlement clusters, based on ethnic or religious ties, scattered across the Prairie Provinces and throughout northern Ontario and Québec, some down to the level of the extended family. Four dozen Norwegian settlers located in the area they named Northfield shortly after it was opened for homestead filing in 1913. A local history, written by one of the "originals" and his wife, provides information on where a homesteader had come from, by what route, and the quarter-section, township, and range on which he located. As revealed by figure 5.6, part of their district is better endowed than most of the case study sites, having access to class 2 and 3 land but qualified by a complex surface. Climate is a major limitation, and on the nearby slopes of the Saddle Hills the soils are classes 4 and 5 with problems of stoniness and high moisture content. The original settlement was made up of people who were born in Norway and came to the southern Peace River country directly or with a stop in the northern United States or central Alberta, or of people born in the latter two locations to immigrant Norwegian parents. The main concentration of Norwegian homesteaders in the Northfield area from about 1913 to 1917, can be mapped by extracting information from

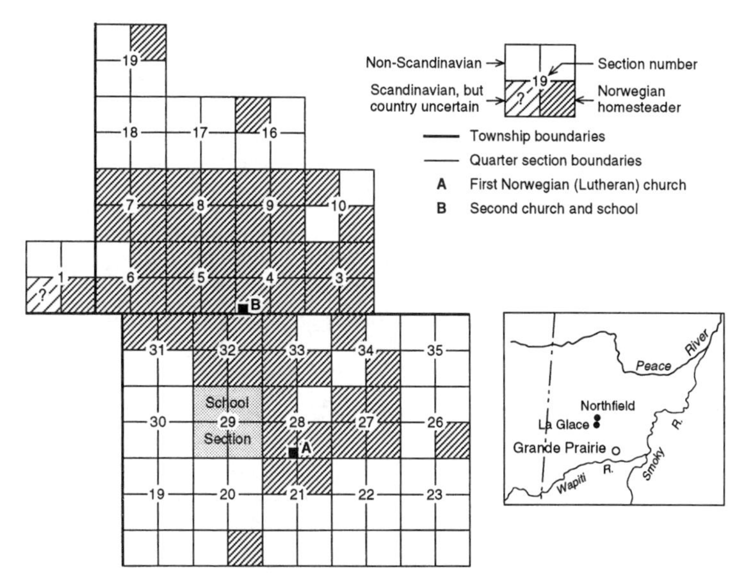


Figure 5.5
Clustering of early settlers and cultural facilities of Norwegian origin in the Northfield area of the South Peace River district, Alberta, around World War I.
*Source:* Interpreted from Johnson, *The Northfield Settlement*, brief family histories, land ownership map, and local community history.

the family stories in the local history book and by building on its map of early landowners (figure 5.5; see also figure 5.6).[71] The map of the Northfield settlement repeats a pattern laid down by ethnic and religious groups in various locations across the Canadian terrain, including elsewhere on the boreal margin: as far as possible, people with common allegiance claim land side by side, they tend to cluster around a church, community hall, or school (even moving such a facility to be more central), and they make a cultural imprint, perhaps by the shape of fields or type of fencing or details and materials of structures. For instance, a church might not be just a church but present a strong cultural declaration, as in a striking onion dome or a four-square, white Mennonite meeting house.

The Northfield area was relatively late in being opened for settlement even for the Peace River country, the southern

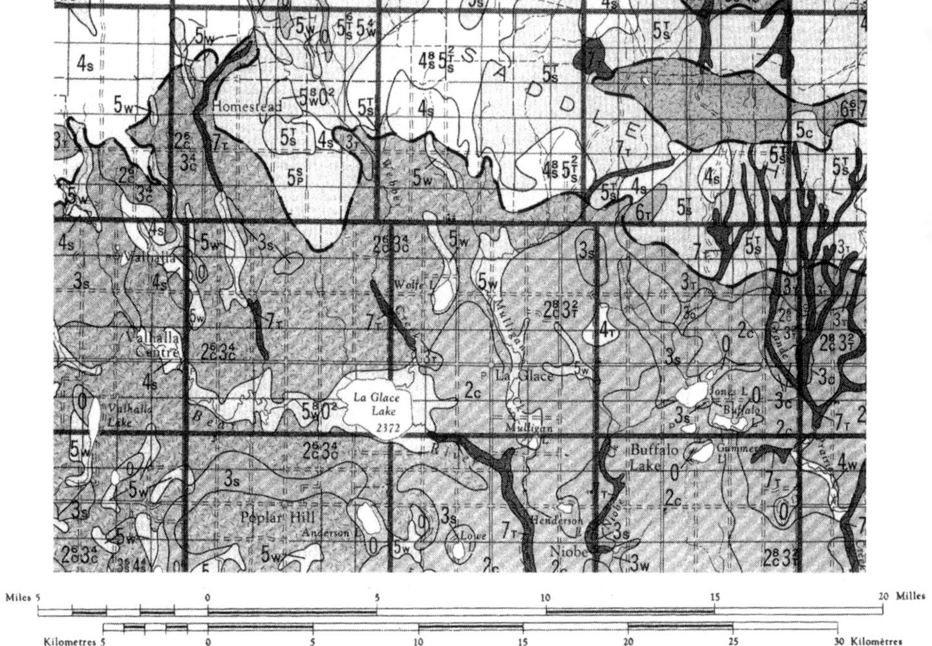


Figure 5.6
CLI assessment of soil quality for agriculture, La Glace-Saddle Hills, Alberta, embracing the location of the Norwegian Northfield settlement of the late 1910s. The symbols for church and school (B on figure 5.5) are in the centre of the figure, 8 km north of La Glace Lake. The slopes of the hills did not offer much land well suited to farming.
*Source:* Part of CLI map 83M "Grande Prairie."

part of which was first surveyed for homesteading in 1907 followed by an inundation of settlers. The land chosen by the Norwegians was mainly north of the highly desirable parkland around Grande Prairie, into more tree cover on the southern slopes of the Saddle Hills. Many of the early settlers combined hunting, trapping, and fishing with clearing trees and expanding their fields for agriculture. During the early stages of clearing, wheat growing was difficult because of common losses to frost. In addition to the usual pioneer cooperation in establishing homesteads, the Norwegian community flourished and expanded through various culture-based activities, such as the Lutheran church, choirs that sang in the Norwegian language, skiing and team sports, interchanges with other Norwegian settlements, and large picnics. The settlement fashioned its environment through a common cultural understanding of land use and livelihood, often expressed in tools, techniques, and folk objects. The scene within the shaded area of the Northfield map would have had echoes of farming areas in Norway around the time of World War I.

## FADING EXPECTATIONS: THE RELIEF LAND SETTLEMENT

Although not group settlements in the way the term is applied in this chapter, the relief settlement or back-to-the-land initiatives of the early 1930s did send out large numbers of primarily urban residents to similar and sometimes contiguous locations. Because the dry margin had proven generally an area of failure, the areas chosen for relief settlement were almost all in or near the boreal margin. As revealed by the commentaries on the relief settlers in the Ontario clay belts, earlier in this chapter, the back-to-the-land schemes were widely believed to be fruitless initiatives and a waste of money.

A report on the relief settlement in Alberta provides a judgment, based on a thorough survey and interviews with seventy of the settlers, and a qualification of the negative view.[72] The land made available for this purpose in Alberta was boreal woodland in the western approaches to the mountains or in the northern extremities of farming country. As the Rackham report summarizes it (a description probably applicable in each of the provinces), "most of the settlers were located in comparatively undeveloped, or completely virgin, grey wooded and transition soil type areas. Since each settler had to acquire his own land there was little tendency to settle in groups, as such, although settlement did seem to concentrate in some areas where there was more land, either suitable, or available, for settlement. In general, settlement was scattered in remote bush districts."[73] An applicant for relief settlement was required to find his own farm property and have it approved by an inspector. Just under 1,100 did so in the ten years that the scheme continued, 1932–41, the peak years by far being 1933 and 1934. The choice of land included unclaimed homesteads (in the last eight years that they were available) (40 per cent), rentals (27 per cent), various kinds of contracts or leases (29 per cent), and personally owned property. A sum of $600 to $700, two-thirds of it in the form of a loan, was given each settler family for purchase of necessary equipment, livestock, and sustenance for up to three years. The three levels of government made a net outlay of just over half a million dollars, in the process saving expenditure from another compartment of the public purse by removing these people from the urban welfare rolls. The calculation, in 1948, of the success of the scheme leaves a large question about the efficacy of the relief settlement. Only 420 settlers were classified as "established," meaning they either had paid off their loan or were supporting themselves on the farm, while the remainder were classified "abandoned." This latter category, however, included a range of reasons, such as enlistment, death, poor health or domestic trouble, other employment, and retirement, not all of which would constitute failure. What we would call out-and-out failure applied fairly to 226, 187 of whom, or 17 per cent of the total, had returned to the relief rolls.[74] The cost to the governments amounted to slightly under $100 per person removed permanently or at least for a period of time from the relief rolls, which was probably less than the cost of having kept them on welfare. Despite this rather sorry calculation, Rackham regards the scheme as having been generally successful, in the sense that a number were established on farms and others were provided with a bridge livelihood before they moved on to different employment.[75]

Another study that bears some comparison with the Alberta example was carried out in Saskatchewan in the 1930s.[76] This example, although contained within a bounded district, was not group-based or for relief purposes. This new settlement was part of the rush to the north, primarily from dried-out areas in the more southerly parts of the province – that is, from prairie to woodland. The time period, the availability of government assistance, and the general environmental conditions were similar to those of the Alberta study, although the 80 km-long strip of townships north of Prince Albert (Albertville to Garrick) was a more concentrated settlement district. The natural vegetation was largely medium to heavy bush except for the western fifth, which was a transition from the parkland to woodland. Soils varied from very fine sand through clays and loams to peat. A random sample of 304 settlers was taken, ranging from some who were in debt when settling to others with two or three thousand dollars of "net worth." Sixty-eight per cent had found homesteads, some in recently surveyed areas, and 22 per cent had bought land. Over the six to twelve years of farm-making in the district all classes of settlers in the sample had increased their average net worth. But, significantly, the conclusion was that "The present level of living ... is considerably below the conception of a reasonable standard ... for a livelihood on Northern Saskatchewan Pioneer Farms."[77] This is without considering any farms that had been abandoned in that decade, during which World War II had begun.

## COMMON EXPERIENCES

Informed opinion in the inter-war period held that group settlement was more stable and quicker to develop infrastructure than settlement by individuals. This opinion was based primarily on examples of group settlements that had been inaugurated by the turn of the twentieth century, most of which had a strong ethnic or religious focus and were early enough to claim good farming territory. These ingredients for success were not available to the post-World War I soldier settlements. Even some earlier group settlements in the boreal belt with an ethnic or religious identity, such as Doukhobor or French Canadian, did not maintain their cohesion. Without doubt living the frontier experience after 1910 was hugely daunting, both physically and psychologically, and probably more comparable to the frontier life in the early nineteenth-century eastern woodlands than to the spread of farming across the prairie and parkland areas. As in the eastern woodlands, clearing land was a much slower process, not only because of the drudgery of chopping trees but also because of the time necessary for assembling the windrows for burning, for finding a safe time to light and tend a fire, and for removing debris and roots. Although power equipment was available in this period for clearing land, few late frontier settlers could even contemplate the expenditure for such assistance.

Apart from the convenience that the group offered, of sharing both jobs and feelings, there were not a lot of differences whatever the form of the settlement on the late frontier. The general condition of the late frontier was isolation, which meant distance for the farmer from markets for farm products, and separation for the individual from most relatives, friends, and medical and dental services. Many social planning experts at the time believed that settlers now had a "quality of life" expectation that previous generations of frontiersmen had not held. Isaiah Bowman included it, in 1927, in the requirements of his "science of settlement": "a new question in pioneering," one of the "questions that nature alone cannot answer, for the questions have to do with the quality of culture that he may enjoy or endure … [i]n the face of the great natural difficulties that the pioneer belts of the world exhibit."[78] Distance continued to be a problem for the northern settlements at least through the first generation, because even as better roads and rail connections came closer, the major markets and city amenities remained in the far distance. In reference to the Swan River valley, Manitoba, Murchie and Grant put it in unadorned language: "The country is promising … to the settler who wishes to make a home, who is willing to rough it a bit, and to forgo many of the customary comforts

... and who above all is willing *for a considerable period to glean his living from his garden and his livestock*" – not what could be described as a glowing endorsement of the boreal margin.[79] There appears to be justification for the complaints of some settlers, especially in government-supported soldier settlement or back-to-the-land projects, that what was offered by "head office" was not always understood or honoured by the agent in the field.[80] Soldier settlers in particular became disillusioned during the 1920s, and for various reasons fled the farms they had been encouraged to enter at the end of the war. But ironically a countervailing stream of people, including the British family scheme participants and Eastern Europeans, was anxious to try where others had failed, so that the whole boreal margin was gaining population, and the Abitibi and Alberta Peace River settlements more than doubled between 1926 and 1931.[81]

Many of the common experiences across the boreal margin raised the spectre of failure in the human endeavour to use nature. The examples investigated above underline the universal significance of climate in the late-settled areas, especially the general coolness, but perhaps more specifically the unpredictable frosts. An extreme climatic event that gained great notoriety through western literature (as it had much earlier through eyewitness accounts from the eastern frontier) was the blizzard. Heavy snow and strong winds could be a lethal combination for even a short-distance traveller, although it appears that this danger loomed much larger on the open prairie, where there were few trees to break the sweep of weather. A widely typical characteristic of the boreal margin that the soil maps reveal is a kaleidoscopic mixture of soil types and landscape that could hide under the blanket of poplar and evergreen. In contrast to the dry margin, the problem in the north often was an excess of water appearing all too regularly across the surface, making clearing difficult and delaying spring access to fields for planting. Climatic characteristics, soil quality, and surface conditions were the primary challenges that faced settlers and sometimes defeated them.[82] Other environmental characteristics could add to the difficulties and threaten the survival of boreal settlement, including periodic outbursts of grasshoppers, especially from the late 1920s, and of wheat stem rust.

Common experiences of a less monumental scale could be significant to people in pioneer conditions. As Dr Percy's letters illustrate, disease was often a problem and, as with the 1918 flu epidemic, premature death was a possibility. Psychological stress, sometimes in the immobilizing form of "cabin fever," could arise from the isolation or the unrelenting environmental conditions or the scale of the challenge of farm-making. Pioneering women often reported, even in the 1920s, that they had not conversed with another woman for months. Unlike earlier eras when off-farm employment was available in railway construction or in towns, distance and more restricted opportunities, especially in the 1930s, deprived settlers of extra income that could sustain the embryonic farm. Settlers in New Ontario, like those in the northern farming areas of the Prairie Provinces, discovered that they could not duplicate the process of Old Ontario, where the clearing of the land for crops led to a rapid increase in the real-estate value of the farm. In New Ontario, once the timber had been sold for lumber or paper pulp, the low value crops that the land was suited for were too far from markets to generate sufficient income. In northern Québec, too, parishes that had been settled during the 1930s survived through "agro-forestry," a well-known marginal way of life divided between self-sufficient agriculture, with a little sale or barter of produce, and money-making timber work that took the men away in the winter season. But by the 1960s the new generation was beginning to abandon that way of life by joining the rural exodus to the towns.[83]

What could be classed as social problems would include the internal dissension that bedevilled some groups, and the discrimination that faced the blacks, eastern Europeans, Doukhobors, Hutterites, and some Jews and Mennonites. On the economic front, farmers faced the apparently uncontrollable divergence between what they could earn and what they had to spend. On the one hand, around the end of World War I costs of farm-making were high, and soldier settlers in particular found themselves quickly in substantial debt for the purchase of land and equipment. On the other hand, the price for the farmer's wheat steadily fell from a high in 1919 to a low in 1929, when many farmers failed to cover their costs in growing the crop. Furthermore, it was calculated that what the farmer sold for one dollar in 1929 brought him fifty-eight cents in January 1932, while what cost him one dollar in 1929 cost ninety cents in 1932.[84] A calculation of farm income compared to farm taxes in Manitoba shows a devastating divergence in the 1920s: the value of farm products sold fell from $165 million in 1920 to $69 million in 1930, while the tax per $100 of assessment went from $2.36 to $3.43 during the same period.[85] Many soldier settlers had their property revalued, resulting in a lower debt, as an aid to keeping their farm operating. In the Great Clay Belt, hopefully labelled New Ontario, soldier settlers who could not continue the struggle with the forest and the climate were provided with passage to some other part of the province where they could secure a different employment; and a few years later, casualties from the relief settlement were rescued from what they thought to be an exile and returned south. These experiences, shared, to one degree or another, among the people who tried to expand the late farming frontier, underlay the exodus that many of them eventually joined.

The attitude behind the encouragement of marginal settlement by authorities in the period 1910 to 1940 was not only that of colonization as promoted in Québec, but also of a kind of internal colonialism on the part of governments: settlers were directed – sometimes transported – to identified locations; they were expected to apply themselves vigorously to settlement tasks; and they were aided to organize themselves according to a normal model of the ruling core area. Where the government saw fit they were incorporated (e.g., in schooling, in the postal system) in the general systems of the country. The settlers were pawns in the attempt to gain new national ecumene.

# 6

# Reflecting on the Expansion into Marginal Lands

Townships ... are open for settlement without differentiation. Many a settler will be misled into taking up unsuitable lands, and the experience ... will be repeated ... of abandoned farms or else a degenerated population.[1]

... 150 filed. I don't think half of those stayed enough to put in residence. And then every year there was more dropped out.[2]

By the 1930s it was widely believed that the only way to introduce successful farming in the marginal areas was through a major investment to "correct" a natural disadvantage in an otherwise promising area. Such investment would usually mean government involvement, and the virtual disappearance of the epic lone, striving homesteader on his 160 acres. A major initiative was the establishment of the federal Prairie Farm Rehabilitation Administration (PFRA) in 1935, with specific interest in the drought problems of the western interior. Its activities were directed primarily at saving and making usable the water that was found on the surface of the dry areas, through subsidizing the construction of dugouts and stockwater dams, and at removing unsuitable land from agriculture to consolidate it into community pastures.[3] On the provincial level, the extension of settlement at the boreal fringe in Manitoba and Saskatchewan required the dredging of drainage canals, as in the Birch River (northeast of Swan River) and Pasquia (at the mouth of the Carrot River) projects in Manitoba, and others in the Carrot River valley in adjacent Saskatchewan.[4] Governments, the civil service, and the public had learnt a great deal, by the end of the thirties, about the potential of the fringes for farming and a modern way of life. Governments recognized that the investment necessary to entice settlers to stretch the frontier a little farther was seldom justified by the long-term results. For potential settlers the large amount of publicity that had appeared in the popular press made clear that success was much less than assured. The onset of war in 1939 allowed – in some cases forced – both settlers and government officials to focus on the expectations and conditions of marginal farm life: was it worth putting money into; was it worth enduring? Whether to go into military service or simply to move toward fresh opportunities, many people left the boreal fringe around this time, joining an exodus from even better farming districts, as others had fled from the dry margin two decades earlier (illustrated by the population graphs in chapter 2). Many of these migrants had not survived a single generation trying to make a farm in the late marginal lands. Are there settlement lessons to be learned from the somewhat desperate expansion of the 1910s to the 1940s?

## LESSONS FOR NEW SETTLEMENT

Isaiah Bowman's science of settlement set down a checklist by which one could measure the likelihood of an embryonic frontier settlement succeeding. The scholars engaged in the Canadian Frontiers of Settlement series, in tune with Bowman, recognized the need for a modern, well-organized approach to settlement in the world's remaining "pioneer belts." By the time the series was appearing, however, most of Canada's boreal margin (outside Québec) had already been invaded through a somewhat chaotic, unscientific process. One recommendation that was not thoroughly implemented, despite wide acceptance, was the prior scientific assessment by government of the areas being offered to settlers, combined with the communication of the assessment results. W.A. Macintosh, in the Canadian Frontiers of Settlement series in the early 1930s, underlined this deficiency, calling it "regrettable that, under the stress of the emergency, this extension of settlement is being attempted largely by a process of trial and error. There is urgent need of definite planning to avoid the mistakes and wastes of past policies ... It is difficult to estimate the loss occasioned by the lack of soil and economic surveys, a lack that may defeat all attempts at permanent settlement."[5] As early as the mid-1920s Bowman called for the involvement of government agencies in studying the natural conditions of candidate areas, assessing crops for appropriateness, investigating the suitability of applicant settlers for the task, ensuring basic modern facilities, and improving access to the areas identified as suitable for new farm settlement.[6] Such an approach would benefit both individual settlers and groups. Québec came closest to having a system for nurturing new settlements, with both church and state providing structure and assistance. Therefore, it is surprising to learn that in places such as Abitibi, where the colonization effort was so fervent, the rural depopulation during a few decades after World War II brought the success rate there down toward the less than 50 per cent common in post-1910 farming expansion in other parts of Canada.[7] It appears that the planning and guiding of the settlement process in Abitibi only extended the rigorous pioneering to two generations rather than the one common in other parts of boreal Canada. Chester Martin, writing in the Canadian Frontiers of Settlement series in the 1930s, condemned the overall results of the free-homestead system as "a truly appalling record of casualties"; and the previous two decades had been the most dismal part of that record.[8]

Many have reflected on the late farming frontier, sometimes in the form of a novel or reminiscence, sometimes in a more formal, structured report, especially of a settlement initiative into which a government has put organizational effort and public money. Thus, if lessons about establishing new settlements are to be learned from Canada's twentieth-century experiments, perhaps the sober assessment of such an initiative is a place to look. The ambitious and well-intentioned settlement for returned soldiers at Kapuskasing was subjected to a commission of inquiry for the government of Ontario (see excerpts from the report in chapter 5). Though in some ways the Kapuskasing colony was untypical in not having an "allegiance" basis, as with a religious or ethnic group, it was a sizable organization with about a hundred settlers, not counting family members. The site was also classic boreal margin, in the cool depths of the Great Clay Belt of northern Ontario, 120 km by rail west of Cochrane. The careful review by the three-man commission of the three years following the settlement's inauguration lays out the reasons for the scheme, the progression of the settling and the difficulties encountered, the complaints voiced by the settlers, and the deficiencies revealed in the plan. The report concludes with a number of recommendations for salvaging the settlement. Analysis of the thirteen-page report can lead to the identification of widely applicable principles for a planned new settlement:

1 Make available a full scientific assessment of the natural conditions of the site.
2 Make the requirements for applicants to the scheme completely clear.
3 Be sure that settlers' hard work and relevant skills are sufficient qualifications for achieving a satisfactory standard of living in the scheme.
4 Do not change the terms of settlement in mid-stream without agreement of the participants.
5 Ensure that the actions and opinions of the scheme overseer are in tune with the declared intentions of the sponsors.
6 Use only an overseer completely familiar with the history and the needs of the type of settlement.
7 Do not import to the settlement replacement management personnel without wide-based negotiation and agreement.

An understanding and application of these principles or lessons would have smoothed the operation of any of the soldier settlement or other government-sponsored schemes in the 1910s to 40s, and it is arguable that group settlements that have succeeded, such as those based on a religious adherence, have progressed under a version of the list above. Surviving examples of successful group settlement in the boreal forest can be found, such as the extreme northerly Mennonite community at La Crete, Alberta, although it has not been without stresses.[9]

In other parts of the northern world a form of rural settlement, based on using the land, has had some success. Kirk H. Stone had a long career of observing and analysing settlements in the boreal forests, especially in Scandinavia and Alaska. In addition to devising a descriptive categorization of the stages of settlement from fully occupied farmland to the scattered occupance of a margin – his "Geographic Measures of Isolation" – he distilled the features of many examples of perennial northern settlement to devise a template for a scientific, planned approach that he called a General Rural Settling Guide. He recommended guidance and control of the process of new fringe settling. There should be detailed locational planning, including an inventory of the area three years in advance; the settlers should be carefully selected, if possible from a source community within 200 km; and a leading farmer should be encouraged to be one of the settlers. The livelihood would be expected to rely on something in addition to farming, such as forestry, but the size of the farms must be large enough to promise success. The settlers should be scheduled to enter at propitious times, and the provision of credit and materials should be controlled by the managers of the settlement. Even with successful establishment, "new settlement should be expected to take at least twenty to thirty years to begin to mature sociologically, psychologically, politically, and economically."[10]

Any attempt to expand rural farm settlement must take into account what Isaiah Bowman realized by the early 1920s, that the individualistic, laisser-faire land grab by nineteenth-century speculators and homesteaders is not appropriate for settling new land today (if it ever was). The only hope for success rests on detailed prior planning, major capital expenditures, and continuing encouragement and support, as the Manitoba and Saskatchewan governments recognized in their post-World War II projects. Despite the enlightenment of the government sponsors, most of these projects have been problematic and of quite limited success. A few pockets of relatively fertile soil have been discovered and entered by groups and individuals since World War II, in northwestern Alberta, around and south of Fort Vermilion, and in a British Columbia extension of the Peace River country, as well as the high-cost, environment-modifying projects in Saskatchewan and Manitoba forming growth points in Vanderhill's northern "ragged edge" of farming.[11]

## WHERE HAVE THE MARGINAL FARMING AREAS GONE?

The failures in the arid margin had generally occurred before the 1930s, and even the alternative expansion into the boreal margin was almost universally losing population by the end of that decade (as illustrated by the population graphs in chapter 2). There is little doubt that in both the dry and the northern fringes plough-farming and especially wheat-farming as the dominant land use had reached, and in some places gone beyond, their limits. Toward the north the late expansion of farm settlement rather quickly met an impregnable boundary almost everywhere. Yet today many parts of the late frontier that were in serious trouble in the 1930s do not look like wasteland. Here and there along the northern margin one sees active individual farms or clusters of farms. The explanation lies in the type of farming: the north was impregnable for wheat and certain traditional field crops, but it has proven amenable to extensive farming, such as animal grazing and, in districts where there is an urban market, to such specialties as dairying. The gainful land use to be seen in stretches of the former marginal lands is not the same as that attempted in the 1920s and 30s, but is nevertheless something we could call farming.

Two bases of farming have changed: farm size and mechanization. These two changes were beginning on individual farms even while the general pre-World War II conditions of depression and failure abounded. Added to this has been an almost universal acceptance of a range of chemicals (and their increasing cost) as a major component of farm practice since World War II. Those who survived in the dry areas have added greatly to the average farm size since the 1930s. Vast acreages run beef cattle at the rate of about fifty acres per head. Wilfrid Eggleston reports on the process underway: his father received "a letter from Axel Mattson, a fellow-homesteader … offering to buy our abandoned quarter section … Axel offered a thousand bushels of wheat … It took Axel eight or nine years to squeeze out the thousand bushels from the sketchy harvests … He made similar deals with other old neighbours and in the end picked up several sections of land" (i.e., several square miles).[12] This process of acquisition of land to increase farm size was beginning even in the 1920s when neighbours sold out and moved away, especially in the drier areas (witness the Egglestons). The notable enlargements of farm machinery, on the other hand, gained momentum after World War II and for some farmers became part of an alternative to abandoning their farms. Very large machines mean that any field work that needs to be done can be accomplished relatively quickly and with few workers. The dryland farmers did not rush to take up the mixed farming that was widely recommended as a solution for drought, although they might have been nagged by uncertainty; wheat continued to be king.[13] In southeast Alberta farmers made a modest move toward diversification, with wheat reducing from 84 per cent to 66 per cent of field crops between 1941 and 1956.[14] At the same time farmers worked toward improving farm practices to retain soil moisture. The summer fallowing that left fields bare and susceptible for long periods to wind erosion was gradually abandoned, in favour of leaving previous crop residue or "trash" on the surface. Eggleston, in his retrospective, believes the return of the dry belt to productive occupance is the result of a shift from dominant grain farming to a combination, called farm-ranching, with a major reliance on beef cattle. Along with this change in approach (which was actually legislated for some parts of southeast Alberta in the 1920s), credit is given to the truck, tractor, and automobile, water conservation and irrigation, higher prices for farm products, and new cultivation techniques.[15] The generally less droughty climate after the 1930s also helped make farming more productive than was possible in the previous two decades.

In the north, too, average farm size increased, though not to the same extent as in the dry areas. Even by the 1930s, new farms in the Albertville-Garrick strip north and east of Prince Albert were clearing and breaking land at the average rate of 5.5 acres per year, and eighteen Peace River farms in Alberta had increased their improved land from 176 to 263 acres, on average, in the dozen years up to 1941 (although, untypically, overall size of farm went down slightly).[16] Eckart Ehlers demonstrates the dramatic increase in farm size in an early settled part of the northern Peace River country, where many farms tripled or quadrupled in size from the original 160 acres (quarter section) during fifty years of settlement. Some farms were over 800 acres by the 1960s. And in a detailed study of a genuinely boreal frontier township settled in the late 1920s (township 86, range 7 west of the 6th principal meridian, near Eureka River), Ehlers found that in thirty-five years the average farm size had grown to nearly three quarter sections.[17]

The quality of the northern woodland soils varied widely, however, and the boreal margin became fragmented in appearance because, with abandonment of land, areas not suitable for crops were allowed to revert to bush or, at best, became community pastures or were leased for grazing. The shortness of the growing season did not change. A mixed agriculture with a major component of meat animals, in contrast to the grain specialty to the south, was typical.

The mixture of production that experts had long encouraged prairie farmers to take up, in an attempt to reduce the exorbitant reliance on wheat, developed naturally in the northern fringes. Zaslow notes that livestock raising and dairying had become a common emphasis in pioneering in Québec, and it appeared to move into the Ontario clay belt with French Canadian settlers.[18] Northern Saskatchewan became characterized, by the 1960s, by coarse grains such as oats and barley, and by uncultivated pasture supporting livestock production (primarily of beef) "typical of much of the northern fringe."[19]

Adaptation to farming in the wooded areas was difficult, especially for farmers who had learned the business in the long-settled parts of the St Lawrence basin or on the grasslands. Even major projects with government backing could founder and fail (as the Kapuskasing Colony had done in a more promising time). Manitoba and Saskatchewan embarked on major projects at the end of the thirties and after World War II. In post-war Manitoba many of the principles of scientific settlement were instituted in the Birch River project, including soil surveys, government-sponsored clearing and breaking of a portion of land, general drainage of the block, preparation of roads, and flexibility in terms of land sale. The settlers were mainly from dried-out farms in Saskatchewan.[20] The Pasquia project, 150 km further north, faced more difficulties, including periodic flooding around the mouth of the Carrot River, followed by some abandonment. Further up the Carrot River in Saskatchewan, some provincially sponsored settlement schemes were tried, especially on behalf of returned military personnel. The Carrot River valley required major drainage work, and almost all these initiatives had difficulty getting properly launched and keeping their population.[21] The Saskatchewan government after the war established a large number of co-operative farms in the boreal belt, primarily in the northeast but also in a few places toward the western boundary of the province. In some cases, co-operation went as far as pooling of equipment, and generally government carried out major drainage works, constructed roads, and offered assistance for buildings. But within a few years the co-operative farms were beginning to split up in favour of individual holdings. Vanderhill's reflection on these attempts to push even farther into the boreal margin was "that we are witnessing the beginning of the end of agricultural settlement in the forest frontier of Manitoba and Saskatchewan."[22] This obituary could be applied across most of the boreal margin by the time World War II veterans had had a few years to test the marginal conditions, only deflected in areas where a major revision of the inter-war farming regimen was instituted.

Improvements in the machines involved in the various aspects of farming have had a major impact on the agricultural territory of the country, including what were the inter-war margins. For one thing they are more reliable, making their purchase a relatively safe alternative to the traditional hiring of farm workers. Machines have also increased in size, thus making it possible to do more work for the same investment of operator time. A great variety of machines have been ingeniously adapted to do certain jobs, or to do them more effectively and with few collateral disadvantages. Conservation-minded tillage techniques have been especially beneficial in dry areas, and have been an integral part of the modification of dryland farming and the move away from summer fallowing.[23] On the boreal margin the increased size and power of machines made the clearing of trees and brush, the excavation of drainage ditches, and the breaking of new land more manageable.

Three conditions were necessary before mechanization could bring success to farmers who managed to survive the crises in the marginal areas before World War II: the farm size had to be sufficient to justify the expenditure on large sophisticated machines; the farmer had to be sure the machinery would not let him down; and the economy had to be favourable to ensure an income sufficient to pay off the huge investment involved in machine purchase. Such conditions did occur after the 1930s, but rather than leading to much expansion of the farmed fringes, it resulted primarily in the consolidation of effort on the better areas. The boreal margin was less suited than the open prairie to the large fieldwork machines. Mechanization got under way in the inter-war period with the gradual shift from horses to tractors. Farms on the prairie were quick to make this move, whereas the farms in the bush country were smaller and took on this new expense more slowly (recall the Kinderwaters' attachment to horses in the Peace River country). But acquisition of machines meant a marked reduction in expenditure for hired labour, and a demonstrable increase in efficiency of the farming enterprise.[24] Perhaps most significant for the northern farmer who could afford a self-binding reaper was the speed with which a grain crop could be harvested in the face of an impending frost.[25] The dry areas in general, with their espousal of modern farming techniques and lighter pressure on the land, no longer look particularly "marginal," excepting the patches fundamentally damaged in the twenties and thirties or poisoned by salinization. The boreal margin, on the other hand, still carries an aura of uncertainty, with decent farms interspersed with indigent areas where the struggle has been given up and the land is reverting to brush and tree cover.

In addition to an increase in beef cattle and a reduction in wheat acreage, innovation in various aspects of the crop complex has allowed large parts of the marginal areas to carry on some productive land use. For one thing, experimentation in modifying the characteristics of cultivars has periodically had a major effect. Perhaps the most famous example was the development of a quicker-maturing variety of wheat, called Marquis, at the beginning of the twentieth century, that was largely responsible for the northward push of agriculture after 1909. Experimentation on wheat continued, especially to combat stem rust and sawfly and to provide better yields in the face of certain climatic problems. By the 1940s Marquis was being replaced because of its susceptibility to rust.[26] Improvements continued in other cultivars, and a variety of different crops were introduced. In some areas flax became important, in others alfalfa or mustard began to be grown for seed and, especially on the northern fringes, barley became a major crop, largely in place of wheat.

MacPherson and Thompson see the conditions during World War II, in contrast to the wheat boom of World War I, as pushing toward diversification (with help from a persistently low world wheat price and direction by the federal government). Diversification proved to be variable across the country, depending on the suitability of an area for producing a specialized crop. Less apparent than diversification, but perhaps more influential in the long run, were wartime control and management of both sides of the food system, at one time setting maximum prices and at another providing price support to balance farmers' increasing costs. This opened new avenues for nurturing a diversifying, more planned farm economy.[27] The involvement of government or quasi-government agencies in stabilizing the prices of farm products, examples of which are the various marketing boards (e.g., Wheat, Milk), has become a standard feature of the agricultural economy, and co-operative organizing by farmers has had a beneficial effect.

The increase in farm size and the widespread adoption of machinery was augmented, especially after World War II, by a growing range of chemicals. Across the farming terrain of Canada, chemicals to kill weeds have become an integral and widely accepted part of farming practice. To a large extent weed killing and mechanization went hand in hand, and even some of the "progressive" soil conservation practices of recent years rely on the killing of weeds by chemicals. Poisons were used to kill grasshoppers in the inter-war period and since, and a great range of insecticides has been developed since World War II. Many of the early "wonder" chemicals for killing weeds and insects have earned a bad name because of toxic side effects and have become a focus of particular concern when human health has been shown to be at risk. But new, improved generations of designed chemicals have regularly appeared, so that farming practices remain largely unmodified and heavily dependent on chemical support. Fertilizer designed in a chemical laboratory is another fundamental ingredient in modern agriculture. Many agriculturalists claim that on the large acreages now farmed, and especially in the marginal areas, production would be uneconomic and farming would not be possible without the artificial fertilizers. Critics counter by noting that the scant initial testing given to most modern, man-made chemicals makes agriculture's reliance on chemicals a "Faustian bargain," in which the price in poisoning is yet to be paid. The cost of chemical solutions, which seems to steadily increase, has been a major determinant of the adapt-or-abandon outcome of farms in the boreal forest.

Out of the welter of results of the expansion of farming into the boreal forest after 1910 a few areas of continued success have appeared. Probably the most clear-cut "find" was the part of the renowned Peace River country surrounding the two parkland cores, one in the south and the other in the north Peace. The parkland had been rapidly settled following the survey that began in 1907. The late expansion beyond the parkland included great stretches especially west and north of Grande Prairie, in the Spirit River district (the central Peace in Alberta), west and north of Fairview in the north Peace, and around Dawson Creek and on the north side of the Peace in British Columbia (see illustrations 5.2 and 5.3). Much of this Peace River country proved very successful in growing high quality grains and specialized seed crops. After the 1920s a few more pockets of good land were found farther and farther to the north in the Peace River basin, including around Notikewin (Dr Percy country), Keg River, and in the environs of Fort Vermilion, where the La Crete clearing was beginning at the end of the thirties. An area in northeastern Saskatchewan, which had been thought of as best suited for forest reserve, was found to be good farming country. This was in a 130 km east-west strip of townships from just north of Prince Albert, reaching along the north side of the Saskatchewan River valley, and petering out in the vicinity of Nipawin. A less substantial pocket of desirable land was found in adjacent Manitoba, an extension north and east of the Swan River district. It included part of the area called Minitonas, included patches of class 2 and 3 land, and stretched for about forty kilometres around the eastern flanks of Porcupine Mountain north of the town of Swan River (see figure 4.2).

Other, more limited areas of farmland were discovered around the time of World War I along the boreal forest fringe. These would include, in addition to the clay belts discussed in detail above, the districts between Rainy River and Fort Frances, west and south of Fort William-Port Arthur (now Thunder Bay), and around Dryden in Ontario (all of which were rapidly shrinking in farm activity by the 1950s).[28] Also land was opened and cleared on the east side of the Smoky River tributary of the Peace River, between the towns of Peace River and High Prairie, led by a large French Canadian influx. More typical of the boreal margin across the country, however, was a scene of rough pasture or hay close to the road, with heavy forest occupying much of the back part of the farms, and with open areas being invaded by bush. This scene was usually the remnant of ambitious attempts to reinvent the wheat frontier, as the Dean of Agriculture at the University of Alberta noted a few years after World War II: "When forested soils are cleared and used solely for grain production, the meagre humus content is rather quickly lowered and yields usually decrease accordingly. Thousands of acres of once cultivated land now abandoned in our forest areas are mute proof of this."[29] Here and there in marginal areas would be a farm or two with better than average conditions that could provide the local district with dairy products and perhaps certain vegetables and feed grain. But apart from the three large, desirable areas described above – the Peace River country, the strip northeast of Prince Albert, and the Birch River area near Swan Lake – most of the boreal fringe that remained open after World War II devolved into a mixed economy in which agricultural activity was combined with, and relied on, income from other sources. The other sources commonly included timber-related jobs, hunting and trapping, working for municipalities, or freighting. This mixed livelihood was not the frontier dream.

Where have the marginal farming areas gone? The search for new farmland faded from the public agenda by the 1950s almost everywhere in Canada. But rather than disappear, the hard-won boreal frontier transmuted into different forms. The two major processes at work were, on one hand, the withdrawal of less suitable land from a farm regimen, allowing trees and weeds to reclaim cleared land; and on the other hand, the adaptation of the farming system to the specific conditions of what had proven to be the better areas. So the boreal margin is still with us, in better areas producing high quality seeds or grain, or supporting flexible mixed farming, in other areas providing dairy products for nearby urban populations or, more commonly, being used to raise animals for meat. Wherever farming has survived on the boreal margin, farm size has increased considerably, while larger and more efficient machinery and the application of science and innovations have made it easier for fewer hands to work the larger acreages (as with farming generally in Canada.)[30] The increase in farm size and the exodus of people from poor farms has left the population widely scattered and the social fabric more threadbare. After World War II, various agricultural improvement associations, with assistance from federal and provincial governments, helped to disseminate scientific information and new practices. Replacing the somewhat indiscriminate push into the northern woodlands that characterized the boreal frontier between the world wars, the choice areas uncovered at that time are now assimilated into Canada's farming domain, while the remainder of the northern fringe has ceased to be of interest for agriculture.

The Great Clay Belt of northern Ontario is a classic example of the futility of much of the hopeful expansion of farming into the boreal forest. In many ways the boreal farming frontier *did* disappear in that once-celebrated district (see illustration 4.1). An array of nearly five dozen townships in Cochrane District, in which some farming was carried on in the 1930s, was reduced to approximately half a dozen by 1976 and, although the average farm size increased, over 70 per cent of the 1941 farmland was lost.[31] Even on the larger farms in the Great Clay Belt the dedication to farming is belied by the record: well under half the farmland was improved in 1992, and only 31 per cent was in crops (similar to the 1941–71 record for the Meadow Lake district in Saskatchewan).[32]

## THE ETHICS OF THE PUSH TO THE MARGINS

The opening of new land for farming after 1910 and the rush of people into that land present a saga of human desire and elusive opportunity. By most measures, the occupation of the marginal areas during the inter-war period was widely blighted by failure. Huge costs were paid both by governments and individuals, and also by charitable agencies during the periodic crises that struck the marginal settlements. Mistakes were made and, although blaming in retrospect is not appropriate, putting on record the character and sources of the errors should provide a useful lesson.

We know that individuals were clamouring for land; for those who understood the situation it was a rather desperate bid to get something decent out of what was left. A similar depth of need was felt by those fleeing to this new land from persecution and loss of property, such as the Amber Valley blacks from Oklahoma and the Mennonites from Russia (some of whom settled in northcentral Saskatchewan and in the Cochrane District of Ontario in the 1920s). Chapter 4 describes other kinds of intentions and needs that were involved in the push into the marginal lands after 1910. In the euphoria of the new century there were many influential voices to reiterate Laurier's claim to the century for Canada. Success apparently lay in perpetuating the activities that had given rise to Laurier's bravado. But it was fairly well known, especially by governing officials with access to the latest science, that the best of the frontier was already occupied by the beginning of the twentieth century. Governments at various levels became committed at one time or another to encouraging people to try newly opened land. Some ministers and bureaucrats in the federal government, who had access to the best information about potential settlement areas, seemed to harbour almost a "manifest destiny" kind of ambition for filling up marginal areas. Government officials at a distance from the challenges of the boreal frontier could afford to romanticize and encourage the expansion of the ecumene. There were signs of consternation over the possibility that the frontier could be coming to an end. Some such apprehension could be found in the provincial governments of Québec, Manitoba, and Saskatchewan, and at times in Ontario and Alberta. This keen interest in expanding the areas of farming appeared most clearly as a solution to the large number of returning servicemen in the late teens and to the armies of unemployed in the early thirties. The catastrophic economic conditions of the 1930s brought out the three levels of government as champions of land settlement through the back-to-the-land or relief settlement initiatives, primarily to remove welfare recipients from the cities.

Another powerful agency in pushing the late frontier was found in the churches. Of course religious affiliation was deeply involved in land settlement in the nineteenth and twentieth centuries in famous cases such as the Mormons, the Mennonites, the Doukhobors, and the Hutterites, of which only the Doukhobors attempted to establish a sizable colony in the northern woodland. But the major precursor of these in Canada had been the Roman Catholic church in Québec, which had been active from the middle of the nineteenth century, gradually creating a system of rural settlement and generating a nationalistic yearning for an agrarian livelihood. The Québécois incursion into northern Ontario seemed to be almost a natural spread from Abitibi and Témiscamingue. The Catholic church made several attempts to establish settlements in the Western provinces, with only modest results. But eventually successful French Canadian settlements were established, with the encouragement of the church, in the boreal margin north of Prince Albert and in the Peace River country northwest of Lesser Slave Lake; and German Catholics had located earlier in a colony northeast of Saskatoon. In a somewhat similar initiative to that of the churches, some nationalist associations were also active in promoting the immigration of their compatriots, usually to neighbouring properties.

The railways clearly had a vested interest in attracting settlers to land near their lines. The character of the soil and climate, and whether the settler seemed headed for success or failure, although of concern to railway officials, were secondary to getting a good price for the property. The proximity of a railway depot was said to be a necessity for success for a farmer, but generally for the settler in the boreal margin the definition of "proximity" was hypothetical for the early years, if not because of difficulty getting to the railway, then because of the expense of shipping long distances. In collaboration with the railways were the transatlantic shipping companies, who also were in the business of funnelling immigrants to whatever might lie in store for them in the interior of the country.

The last spasm of the agricultural frontier in Canada, from about 1910, shunted often naive settlers into what were generally less promising conditions than had been the lot of farmers in the St Lawrence valley and the Prairie Provinces. The dry margin revealed its true nature even before the end of the 1910s, and population was leaving it in significant numbers ten years before the major drought of the thirties (as seen in figure 2.2). The boreal margin had been slower to open for settlement, primarily because of the well-known challenge of clearing woodland, but for many of the "dried-out" settlers leaving arid areas it became a refuge. The northern woodland, however, was a mixed blessing, and for some unfortunate settlers failure elsewhere was followed by failure here, brought on by too many frosts, or wetness, or low productivity. By the 1930s this version of the frontier was not engendering the positive mentality produced by the nineteenth-century frontier. In retrospect it is obvious that it could not. In most respects, and in most of its boreal extent, the late frontier was at its limits, and the realization of this by thoughtful observers came as a shock – a fatal blow to the optimism that flowed from the expansion of the nineteenth-century frontier.

The ethical issue that stands unresolved in the record of the settling of the late margins in Canada is the apportionment of blame for the waste and suffering that resulted, whether from individual incompetence, the colonization zeal of churches and other agencies, or the kind of internal colonialism administered by various governments. Contemplating this issue might at least engender caution and provide a perspective that could be applied to related social administration. Indeed, the sad results of the inter-war new settlement proved to have some influence on planning after World War II. We are close now to having a "science of settlement," including tools like the Canada Land Inventory of land for agriculture (shown in chapter 5) as well as sophisticated climatic measurements, and governments are prepared to implement it. But eligible land is no longer in abundance. Perhaps there is also understanding of what was in the 1910s, and even in the twenties, a national mentality that failed to see through nineteenth-century optimism to the reality of the margins. The juggernaut of the frontier legend and the inspiration of having occupied a continent from sea to sea fettered the judgment and raised the hopes of all kinds of people who wanted to see new land opened. Scientific knowledge, or what might be termed geographical intelligence, was available to the government officials who encouraged the occupation of the marginal lands, but naming the limitations of the late boreal frontier and recognizing the end of the era of land expansion were steps they would not take in time to save the many benighted settlers.

APPENDIX A

# The Frontier as Male Territory

A Sampling of the Ratio of Males to 100 Females in a Selection of Boreal Forest Settlement Areas, 1911–31, with Figures for Provinces as Context.

| *Census unit* | *1911 (or 16)* | *1921* | *1931* |
|---|---|---|---|
| QUEBEC | 102 | 107 (rural) | 110 (rural) |
| Abitibi District | – | 149 (age 15+)<br>134 (rural) | 149 (15+)<br>131 (rural) |
| Senneterre E & W | – | 164 | 142 |
| La Motte W | – | 126 | 105 |
| Fiedmont/Barraute | – | 160 | 139 |
| Poularies | – | 150 | 133 |
| Roquemaure | – | (Palmarolle: 33 m. 11 f.) | 110 |
| ONTARIO | 106 | 113 (rural) | 117 (rural) |
| Cochrane District | – | 180 (15+)<br>165 (rural) | 169 (rural) |
| Clergue | 162 | 123 | 130 |
| Taylor | 168 | 137 | 123 |
| Glackmeyer | 162 | 122 | 133 |
| Shackleton | – | 153 | 127 |
| Lowther | – | 238 | 221 |
| MANITOBA | 122 | 119 (rural) | 119 (rural) |
| Division 15 | – | 129 (rural) | 129 (15+)<br>124 (rural) |
| Minitonas District | 134 (16) | 136 | – |
| Tp37.Ra24.W1 | 11 m. 5 f. | NA | 154 |

| Census unit | 1911 (or 16) | 1921 | 1931 |
|---|---|---|---|
| Tp38.Ra25.W1 | 15 m. 5 f. | NA | 111 |
| Swan River District | 126 (16) | 125 | – |
| Tp37.Ra25.W1 | 11 m. 12 f. | NA | 113 |
| Tp35.Ra25.W1 | 8 m. 6 f. | NA | 113 |
| Tp40.Ra26.W1 (Div. 16) | – | – | 132 |
| SASKATCHEWAN | 145 | 130 (rural) | 125 (rural) |
| Div. 3 (dry margin) | – | 128 (rural) | 131 (15+)<br>120 (rural) |
| Div. 8 (dry margin) | – | 128 (rural) | 136 (15+)<br>125 (rural) |
| Div. 14 | – | 136 (rural) | 150 (15+)<br>133 (rural) |
| Tp40–2.Ra10–12.W2 | 123 (16) | 181 | – |
| Tp52–4.Ra10–14.W2 | 8 m. 0 f. (16) | 8 m. 3 f. | – |
| Tp40.Ra9.W2 | – | – | 129 |
| Tp42.Ra7.W2 | – | – | 180 |
| Tp53.Ra14.W2 | – | – | 200 |
| Div. 15 | – | 117 (rural) | 130 (15+)<br>123 (rural) |
| Tp52–4.Ra25–8.W2 | 208 (16) | 149 | 144 |
| Tp52–4.Ra22–4.W2 | 28 m. 8 f. (16) | 181 | 162 |
| Tp51.Ra20.W2 | – | – | 157 |
| Tp52.Ra24.W2 | – | – | 197 |
| Tp53.Ra25.W2 | – | – | 126 |
| Div. 16 | – | 125 (rural) | 143 (15+)<br>133 (rural) |
| Tp52–4.Ra7–9.W3 | 148 (16) | 132 | 194 |
| Tp52.Ra8.W3 | 13 m. 14 f. | NA | 115 |
| Div. 17 | – | 131 (rural) | 145 (15+)<br>131 (rural) |
| Tp56–8.Ra19–21.W3 | 20 m. 25 f. (16) | 130 | 175 |
| Tp55.Ra22.W3 | 3 m. 0 f. | NA | 118 |
| Tp59.Ra18.W3 | – | – | 136 |
| ALBERTA | 149 | 134 (rural) | 131 (rural) |
| Div. 1 (dry margin) | – | 133 (rural) | 130 (15+)<br>136 (rural) |
| Div. 3 (dry margin) | – | 145 (rural) | 158 (15+)<br>140 (rural) |
| Div. 9 | – | 137 (rural) | 155 (15+)<br>140 (rural) |
| Tp44.Ra2.W5 | – | – | 111 |

| *Census unit* | *1911 (or 16)* | *1921* | *1931* |
|---|---|---|---|
| Div. 13 | – | 123 (rural) | 140 (15+)<br>124 (rural) |
| Tp62.Ra2.W4 | 16 m. 7 f. | NA | 217 |
| Tp63–6.Ra1–3.W4 | 193 (16) | – | – |
| Div. 14 | – | 133 (rural) | 147 (15+)<br>129 (rural) |
| Tp64–5.Ra19–21.W4 | 155 (16) | – | – |
| Div. 16 | – | 166 (rural) | 170 (15+)<br>145 (rural) |
| Tp72–4.Ra11–13.W6 | 26 m. 12 f. (16) | NA | NA |
| Tp84–6.Ra5–7.W6 | 5 m. 2 f. (16) | – | – |
| Tp74.Ra13.W6 | – | – | 155 |
| Tp86.Ra5.W6 | – | – | 124 |

*Notes:*
The figures most revealing of conditions in the boreal margin are for *adults* and *rural areas*, but they were rarely available for smaller units in the census. Note that in some censuses, returns for smaller units were amalgamated (e.g., Tp40–2.Ra10–12.w2 [the combining of 9 townships]; or Minitonas District, a group of unincorporated townships). The lack of a record (usually indicating no residents) under the entity as named is shown by a dash; NA means "not available" in a useful form.

In Québec and Ontario the smallest census unit is the named parish or township, whereas in the Prairie Provinces the smallest unit is the numbered township in the square survey. The township (Tp) in the above table is identified by its number and its range (Ra) west of one of the six principal meridians (e.g., W2) used in the survey (see the principal meridians identified by roman numerals on map figure 1.2). A township was 6 by 6 miles (9.6 × 9.6 km). Townships were numbered south to north, beginning at the U.S. border, and ranges (or columns) were numbered east to west from each principal meridian.

It should be recognized that, especially at the beginning of settlement, population numbers were often exceedingly small (as demonstrated in places in the table; e.g., 11 m[ales] 5 f[emales]). Only where the population in a township numbered more than three dozen is the ratio of males per 100 females used. The small units in this table will also be found on the graphs of total population change. The designation "Local Improvement District," which usually embraced a number of townships, was not used for data because of the difficulty of relating a specific township(s) to it. The variability of the information for small units in the

Prairie Provinces, especially in the 1920s, is partly explained in the appendix to the 1926 census for Saskatchewan: "There being no fixed political divisions ... corresponding to counties in Eastern Canada, it has been found impractical to compare, for rural parts, the data collected at any census with that of a previous one for smaller areas than the province as a whole." Thus to provide some more detailed information, the census bureau divided the province into seventeen "census divisions" (used in the above table), but below that level the returns for smaller units were amalgamated.
*Sources*: Canada, *Census 1911* vol. I, table 1; *Census of the Prairie Provinces, Population and Agriculture 1916*, table IV; *Census 1921* vol. I, tables 16, 20; *Census 1926*, vol. I, 296; *Census 1931*, vol. II, tables 16, 24; *Census of the Prairie Provinces 1936*, vol. I, table 6.

## APPENDIX B

# The Canada Land Inventory Maps

The Canada Land Inventory maps of suitability of land for agriculture were prepared by Canada, Department of Environment, Lands Directorate, in the 1970s and 80s. Most of Canada south of latitude 56° has been mapped. The published maps are in colour. The grey scale used here reduces legibility but does reveal the most important aspect of the boreal margin for settlement – the striking mixture of types and qualities of surface conditions.

The information on the maps can be deciphered by understanding that the larger numerals refer to the land classes (1 to 7), whereas the smaller superscript numbers indicate the tenths of the area attributable to the class. For example $5^5$ $3^3$ $7^2$ shows that 5 tenths of the area is class 5 soil, 3 tenths is class 3, and 2 tenths is class 7. Only soils ranked 1 to 3 would be considered good for agriculture. In many cases additional small letters are inserted below the superscript numbers, indicating some important qualification to the soil ranking. For example, C shows "adverse climate," mainly low temperatures or unreliable rainfall; D, "undesirable soil structure" (e.g., impermeability); P, "stoniness"; R, "shallowness to solid bedrock"; M, "low moisture-holding capability";

W, "excess water other than from flooding"; T, "adverse topography – either steepness or the pattern of slopes limits agricultural use." These so-called "subclass" letters can appear with even the top ranked soils; e.g., class 2 soil for agriculture, but qualified by C indicating agriculture-limiting cold temperatures, a common condition along the boreal frontier.

# Notes

## PREFACE

1 Boulding, Kenneth E., "The Economics of the Coming Spaceship Earth," in *Beyond Economics. Essays on Society, Religion, and Ethics*, Ann Arbor: University of Michigan Press, 1968, 275.
2 Pollard, Sidney, *Marginal Europe. The Contribution of Marginal Lands since the Middle Ages*, Oxford: Clarendon Press, 1997.

## CHAPTER ONE

1 Letter of 20 June 1918, from Saskatchewan Deputy Minister of Agriculture Auld to A.E. MacNab, Secretary of Bruce Preparedness League, Walkerton, Ont., in Saskatchewan Archives Board, Regina: R-261, Department of Agriculture, Deputy Minister files, 26.5 – Soldier Settlements (italics mine).
2 Ontario, Department of Agriculture, Zavitz, E.J., *Report on the Reforestation of Waste Lands in Southern Ontario, 1908*, Toronto: King's Printer, 1909.
3 Séguin, Normand, "L'histoire de l'agriculture et de la colonisation au Québec depuis 1850," and Faucher, Albert,

"Explication socio-économique des migrations dans l'histoire du Québec," in Normand Séguin, ed., *Agriculture et Colonisation au Québec: Aspects historiques*, Montréal: Boréal Express, 1980, 9–37 and 141–58.

4 Tyman, John Langton, *By Section, Township and Range. Studies in Prairie Settlement*, Brandon: Assiniboine Historical Society, 1972, 98–103 (especially fig. 36).

5 Friesen, Gerald, *The Canadian Prairies. A History*, Toronto: University of Toronto Press, 1984, 309–10.

6 Dawson, Carl, and Eva Younge, *Pioneering in the Prairie Provinces: the Social Side of the Settlement Process*, Canadian Frontiers of Settlement, vol. 8, Toronto: Macmillan, 1940.

7 Taylor, Griffith, "The Frontiers of Settlement in Australia," *Geographical Review*, 16 (1926), 1–6; and Powell, J.M., *Griffith Taylor and "Australia Unlimited,"* J.M. Macrossan Memorial Lecture, 13 May 1992, St Lucia: University of Queensland Press, 1993, 19–21.

8 Laut, Agnes C., "The Last Trek to the Last Frontier. The American Settler in the Canadian Northwest," *Century Illustrated Monthly Magazine*, vol. 78 (1909), 102–3.

9 A publication title that captures the essence of marginal farming throughout eastern Canada is *"They Farmed, Among Other Things." Three Cape Breton Case Studies*, by Pieter J. De Vries and Georgina MacNab-De Vries, Sydney, NS: University College of Cape Breton Press, 1983.

10 Tuttle, William M., Jr, "Forerunners of Frederick Jackson Turner: Nineteenth Century British Conservatives and the Frontier Thesis," *Agricultural History*, 41 (1967), 219–27; re Russian precursors, see Savage, William W., Jr, and Stephen I. Thompson, "The Comparative Study of the Frontier: An Introduction," in William W. Savage and Stephen I. Thompson, eds, *The Frontier. Comparative Studies*, vol. 2, Norman: University of Oklahoma Press, 1979, 12.

11 Turner, F.J., "The Significance of the Frontier in American History," in *Frontier and Section. Selected Essays of Frederick Jackson Turner*, Englewood Cliffs: Prentice-Hall, 1961, 37–62; Nelles, H.V., *The Politics of Development. Forests, Mines & Hydro-electric Power in Ontario, 1849–1941*, Toronto:

Macmillan of Canada, 1974, 43. Nelles posits that in northern Ontario (not the West), although farming was a component in development, the provincial government played a larger role in diversifying the economy than was the case on the U.S. frontier.

12 Peale, Norman Vincent, *The Power of Positive Thinking for Young People*, Englewood Cliffs: Prentice-Hall, 1952, 214.

13 Inge, William, "The Idea of Progress," The Romanes Lecture, Oxford: Clarendon Press, 1920; Beard, Charles A., "The Idea of Progress," in Charles A. Beard, ed., *A Century of Progress*, New York: Harper, 1932, 3.

14 Bowman, Isaiah, "The Scientific Study of Settlement," *Geographical Review*, 16 (1926), 647–53; for a fuller version, see his "The Pioneer Fringe," *Foreign Affairs*, 6 (1927), 49–66.

15 "Memorandum for Dr. Bowman, Chairman Committee on Pioneer Belts," 31 July 1926, by O.E. Baker, in University of Saskatchewan Archives, Saskatoon: College of Agriculture, Dean's Correspondence – Canadian Pioneer Belts Project. 1928–31. This same memo and others bearing on the roles of Baker and Bowman are found in the archives of the American Geographical Society, specifically the file "Publications. Pioneer Belts Studies. Committee on Pioneer Belts" and the file "Publications. Pioneer Belts Studies. Meeting at Hanover, 1927."

16 Sources include Bowman, "The Pioneer Fringe," and Bowman's "Introduction," in American Co-ordinating Committee for International Studies, *Limits of Land Settlement. A Preliminary Report on Present-day Possibilities*, U.S. Memorandum No. 2, 1937.

17 Bowen, William A., "Mapping an American Frontier: Oregon in 1850," Map Supplement No. 18, *Annals of the Association of American Geographers*, vol. 65 (Mar. 1975).

18 Baker, O.E., "Government Research in Aid of Settlers and Farmers in the Northern Great Plains States of the United States," in W.L.G. Joerg, ed., *Pioneer Settlement. Cooperative Studies by Twenty-Six Authors*, New York: American Geographical Society, Special Publication No. 14, 1932, 61.

19 Georgeson, C.C., "The Possibilities of Agricultural Settlement in Alaska," in Joerg, ed., *Pioneer Settlement*, 51, 54.
20 Schurz, W.L., "Conditions Affecting Settlement on the Matto Grosso Highland and in the Gran Chaco," Wellington, John H., "Pioneer Settlement in the Union of South Africa," Voshchinin, V.P., "History, Present Policies, and Organization of Internal Colonization in the USSR," Roberts, Stephen H., "History of the Pioneer Fringes in Australia," in Joerg, ed., *Pioneer Settlement*, 108–15, 146–53, 264–9, 397–404. Further studies of expansion that continued through the 1930s can be found in David Harry Miller and Jerome O. Steffen, eds, *The Frontier. Comparative Studies*, Norman: University of Oklahoma Press, 1977.
21 Taylor, "The Frontiers," 5.
22 "Growing degree days" are the total accumulated degrees above 5°C, from daily records for an average year. The 5°C threshold is the approximate temperature at which farm crops start to grow. A day on which the temperature reaches 20°C would yield 15 growing degree days. Canada, Department of Energy, Mines and Resources, *The National Atlas of Canada*, 5th ed., "Growing Degree-Days" map, Ottawa: Geographical Services, Directorate, 1981 (based on 1941–70 records).
23 Fernow, B.E., *Conditions in the Clay Belt of New Ontario*, Ottawa: Commission of Conservation, 1912, 4, 10–11.
24 Fernow, *Conditions*, 11; bold in original.
25 Canada, House of Commons, *Official Report of the Debates*, 5 Edw. VII (1905), vol. I, Ottawa: King's Printer, 1905, 1432–3 passim, re western dissatisfaction.
26 Eggleston, Wilfrid, "The Old Homestead: Romance and Reality," in Howard Palmer, ed., *The Settlement of the West*, Calgary: University of Calgary, Comprint Publishing Co., 1977, 122.
27 Handwritten note by Albright in the 1920s, at the Beaverlodge Agricultural Experimental Station, in Alberta.
28 Albright, W.D., "An Economic Pioneer Land Settlement Policy," CSTA *Review*, no. 35 (Dec. 1942), 14.

29 Powell, Joseph M., *An Historical Geography of Modern Australia: The restive fringe*, Cambridge: Cambridge University Press, 1988; Michael Williams, *The Making of the South Australian Landscape: A Study in the Historical Geography of Australia*, London, New York: Academic Press, 1974; Marilyn Lake, *The Limits of Hope: Soldier Settlement in Victoria, 1915–38*, Melbourne: Oxford University Press, 1987; David Wood, "Limits reaffirmed: new wheat frontiers in Australia, 1916–39," *Journal of Historical Geography*, 23 (1997), 459–77.

30 Roche, Michael, "Fit Land for Heroes: A Reappraisal of Discharged Soldier Settlement in New Zealand after World War I," paper presented to the International Conference of Historical Geographers, Université Laval, Ste Foy, Québec, August 2001; Roche, M., "Soldier Settlement in New Zealand after World War I: two case studies," *New Zealand Geographer* 58 (2002), 23–32; Powell, Joseph M., "The debt of honour: soldier settlement in the Dominions, 1915–1940," *Journal of Australian Studies*, 8 (1981), 64–87.

31 The Guelph Conference on the Conservation of the Natural Resources of Ontario, *Conservation and Post-War Rehabilitation*, [n.p.], 1942, 3, 11.

32 Guelph Conference, *Conservation*, 10.

33 Marsh, George Perkins, *Man and Nature; or, Physical Geography as Modified by Human Action*, [1864], ed., David Lowenthal, Cambridge, MA: Belknap Press, 1965; Carson, Rachel, *Silent Spring*, [1962], Harmondsworth: Penguin, 1965.

34 Marsh, *Man and Nature*, xxiii.

35 Girard, Michel F., *L'écologisme rétrouvé. Essor et déclin de la Commission de la conservation du Canada*, Ottawa: Les Presses de l'Université d'Ottawa, 1994.

36 For precursors see Glacken, Clarence J., *Traces on the Rhodian Shore. Nature and Culture in Western Thought from Ancient Times to the End of the Eighteenth Century*, Berkeley & Los Angeles: University of California Press, 1967.

37 Jones, David C., *Empire of Dust: settling and abandoning the Prairie dry belt*, Edmonton: The University of Alberta Press, 1987.

## CHAPTER TWO

1 "Not Adapted as Settlers," editorial, *Sudbury Star*, 29 Apr. 1933.
2 "La Femme du Colon" (reprinted from *La Canadienne* magazine), *L'Abitibi*, 22 Apr. 1920.
3 England, Robert, *The Colonization of Western Canada. A Study of Contemporary Land Settlement (1896–1934)*, London: King & Son, 1936, 86–7.
4 Osborne, Brian S., and Susan Wurtele, "The Other Railway: Canadian National's Department of Colonization and Agriculture," *Prairie Forum*, 20 (1995), 231–53.
5 Canada, *Census of 1941*, vol. 1, chapter 6, 166.
6 Dawson, C.A., *Group Settlement. Ethnic Communities in Western Canada*, vol. 7 of the Canadian Frontiers of Settlement series, Toronto: Macmillan, 1936; Anderson, Alan B., "Ethnic Identity in Saskatchewan Bloc Settlements: A Sociological Appraisal," in Howard Palmer, ed., *The Settlement of the West*, Calgary: University of Calgary, 1977, 187–225.
7 Canada, *Census of 1941*, vol. 2, table 46; England, *Colonization*, 87–8.
8 Canada, *Census of 1941*, vol. 1, 166.
9 Canada, *Census of 1931*, vol. 2, table 46.
10 In Lower, A.R.M., *Settlement and the Forest Frontier in Eastern Canada*, first part of Canadian Frontiers of Settlement, vol. 9, Toronto: Macmillan, 1936, appendix B. For wider discussion: Wynn, Graeme, *Timber Colony: A Historical Geography of Early Nineteenth Century New Brunswick*, Toronto: University of Toronto Press, 1981, 76–83; also A. Collins, testimony, Saskatchewan, Royal Commission on Immigration and Settlement, 1930, typescript vol. 6, Nipawin, 19 Feb. 1930: "A lot of homesteaders used to work in the lumber camps ... [for] money in the wintertime," Saskatchewan Archives Board, Saskatoon, Dr William Swanson (Chair) papers, A 4 Box 1, 29.
11 Lower, *Settlement and the Forest Frontier*, 149. See the thorough study by W. Robert Wightman and Nancy M.

Wightman, *The Land Between: Northwestern Ontario Resource Development, 1800 to the 1930s*, Toronto: University of Toronto Press, 1997, especially chapter 3.

12 This paragraph is based on data in Canada, *Census of 1931*, vol. 2, table 46.

13 Collins, testimony, Saskatchewan, Royal Commission on Immigration and Settlement, vol. 6, 24, 28.

14 Danysk, Cecilia, *Hired Hands: Labour and the Development of Prairie Agriculture, 1880–1930*, Toronto: McClelland and Stewart, 1995, 64–76; *idem.*, "'A Bachelor's Paradise': Homesteaders, Hired Hands, and the Construction of Masculinity, 1880–1930," in Catherine Cavanaugh and Jeremy Mouat, eds, *Making Western Canada. Essays on European Colonization and Settlement*, Toronto: Garamond Press, 1996, 154–85.

15 Langford, Nanci L., "First Generation and Lasting Impressions: The Gendered Identities of Prairie Homestead Women," PhD thesis, University of Alberta, 1994.

16 Letter from Mrs Peter Stenhouse to Saskatchewan Minister of Agriculture, 3 Apr. 1936, in Saskatchewan Department of Agriculture, R-261, 2.229, Saskatchewan Archives Board, Regina.

17 Letter from T.D. Cunningham, Morinville, AB, to Premier Aberhart, 18 Nov. 1936, in Premiers' Papers, file 2280A, Alberta Provincial Archives.

18 Mrs W.H. Howes, in transcript of tape-recorded interview with Mr and Mrs Howes, Saskatchewan Archives Board, C. 19, 23.

19 Minute book of Shaunavon District Home Makers Club, 16 Sept. 1937, Saskatchewan Archives Board, Regina, R-1151.

20 Canada, *Census of 1931*, vol. 12, table 2a, 805.

21 Dawson, C.A., "The Social Structure of a Pioneer Area as Illustrated by the Peace River District," in Joerg, ed., *Pioneer Settlement*, 45–6.

22 Dawson and Younge, *Pioneering*, 311–14.

23 McInnis, R.M., "Childbearing and Land Availability: Some Evidence from Individual Household Data," in Ronald D. Lee, ed., *Population Patterns in the Past*, New York: Academic Press, 1977, 201–27.

24 The usual break in the age range in the township clerks' annual reports up to 1840 (16). Data are in the Archives of Ontario, Municipal Records, RG 21. Further elaboration: Wood, J. David, "The Population of Ontario: A Study of the Foundation of a Social Geography," in Guy M. Robinson, ed., *A Social Geography of Canada,* Toronto: Dundurn Press, 1991, especially 99–101. Also see Easterlin, Richard A., "Population Change and Farm Settlement in the Northern United States," *Journal of Economic History,* 36 (Mar. 1976), 45–75.
25 Canada, *Census of Canada 1931*, vol. 2, table 24.
26 Barron, Hal S., *Those Who Stayed Behind: Rural Society in Nineteenth-Century New England*, Cambridge: Cambridge University Press, 1984, 89; Curti, Merle, *The Making of An American Community: A Case Study of Democracy in a Frontier County,* Stanford: Stanford University Press, 1959, 68.
27 Voisey, Paul, *Vulcan: The Making of a Prairie Community,* Toronto: University of Toronto Press, 1988, 33–6.
28 Voisey, *Vulcan*, 34–5, fig. 4.
29 Wood, J. David, "Population Change on an Agricultural Frontier: Upper Canada, 1796 to 1841," in Roger Hall, W. Westfall, L.S. MacDowell, eds, *Patterns of the Past. Interpreting Ontario's History,* Toronto: Dundurn Press, 1988, 55–77; Norris, Darrell, "Household and Transiency in a Loyalist Township: the People of Adolphustown, 1784–1822," *Histoire sociale/Social History,* 13, no. 26 (1980), 399–415; Gagan, David, *Hopeful Travellers: Families, Land, and Social Change in Mid-Victorian Peel County, Canada West*, Toronto: University of Toronto Press, 1980.
30 Gagan, David, and H. Mays, "Historical Demography and Canadian Social History: Families and Land in Peel County, Ontario," *Canadian Historical Review,* 54 (1973), 27–47.
31 Ost, Lillian, ed., *Seven Persons. One Hundred Sixty Acres and a Dream*, Medicine Hat: Seven Persons Historical Society, 1982, 46–7.
32 Séguin, Normand, ed., *Agriculture et Colonisation*; Courville, Serge, "Des Campagnes Inachevées: L'Exemple du Nord Québécois," in Brian S. Osborne, ed., *Canada's Countryside* (forthcoming). For an international survey, see Fedorowich,

Kent, *Unfit for heroes. Reconstruction and Soldier Settlement in the Empire between the Wars*, Manchester: Manchester University Press, 1995.

33 Fernow, *Conditions.*

34 Bouchard, Gérard, "Family Reproduction in New Rural Areas: Outline of a North American Model," *Canadian Historical Review*, 74 (1994), 475–510.

35 Badè, William Frederic, *The Life and Letters of John Muir*, Boston: Houghton Mifflin, 1924, vol. 1, 142.

36 Canada, *Census of 1931*, vol. 2, table 16; Canada, *Census of 1941*, vol. II, table 13.

37 Canada, *Census of 1931*, vol. 2, table 16; Canada, *Census of 1941*, vol. II, table 13.

38 Canada, *Census of Alberta, 1936*, table 14

39 Clark, Bertha W., "The Huterian Communities," *Journal of Political Economy*, 32 (1924), 357–74, 468–86.

40 Tracie, Carl J., *"Toil and Peaceful Life." Doukhobor Village Settlement in Saskatchewan, 1899–1918*, Regina: Canadian Plains Research Centre, University of Regina, 1996; Katz, Yossi, and John Lehr, "Jewish and Mormon Agricultural Settlement in Western Canada: a Comparative Analysis," *Canadian Geographer*, 35, no. 2 (1991), 128–42.

41 The map can be seen as the frontispiece to Dawson, *Group Settlement*. England, Robert, *The Central European Immigrant in Canada*, Toronto: Macmillan, 1929; Grove, Frederick P., *Settlers of the Marsh*, New Canadian Library, no. 50, Toronto: McClelland & Stewart, 1966.

42 Semple, Neil, *The Lord's Dominion: The History of Methodism in Canada*, Montreal-Kingston: McGill-Queen's University Press, 1996, 216, 224, 349.

43 Cloke, Paul, and Jo Little, eds, *Contested Countryside Cultures: Otherness, Marginalisation and Rurality*, London: Routledge, 1997; Shields, Rob, *Places on the Margin: Alternative Geographies of Modernity*, London: Routledge, 1991.

44 Lower, *Settlement and the Forest Frontier*, appendix C.

45 Bouchard, "Family Reproduction," 475–510.

46 Quoted in *The Globe*, 18 April 1935, 11, in Sinclair, Peter W., "Agricultural Colonization in Ontario and Québec: Some

Evidence from the Great Clay Belt, 1900–45," *Canadian Papers in Rural History*, V, D.H. Akenson, ed., Gananoque, ON: Langdale Press, 1986, 115.

47 "Le Climat Abitibien," *La Terre de Chez Nous*, 20 Nov. 1935 (my translation).

48 "La Femme du Colon," *L'Abitibi*, 22 April 1920 (my translation).

49 *Cochrane Northland Post*, 9 Sept. 1921, 7.

50 Cunningham, T.D., letter to Premier William Aberhart, 18 Nov. 1936.

51 Alberta, Executive Council, Orders in Council 641/34, in Provincial Archives of Alberta, Acc. No. 70.427 (1934).

52 Report by Jacob Toews, in Ontario Archives, Premier Howard Ferguson – Correspondence received [1926], RG 03–06–0–1289 (reel MS 1721).

## CHAPTER THREE

1 Murchie, R.W., and H.C. Grant, *Unused Lands of Manitoba. Report of Survey 1926*, Winnipeg: Manitoba Department of Agriculture and Immigration, 1927, 191.

2 Compare with H.J. Fleure's classic depiction of Regions of Effort and Regions of Difficulty in "Human Regions," *Scottish Geographical Magazine*, vol. 35 (1919), 94–105.

3 Rae, Thomas I., *The Administration of the Scottish Frontier: 1513–1603*, Edinburgh: Edinburgh University Press, 1966; Pollard, Sidney, *Marginal Europe. The Contribution of Marginal Lands since the Middle Ages*, Oxford: Clarendon Press, 1997; various essays in Heikki Jussila, Walter Leimgruber, and Roser Majoral, eds, *Perceptions of Marginality. Theoretical issues and regional perceptions of marginality in geographical space*, Aldershot: Ashgate Publishing Ltd, 1998.

4 Reitel, François, "Le Rôle de l'Armée dans la Conservation des Forêts en France," *Bulletin de l'Association de Géographes Français*, no. 502–3 (1984), 143–54.

5 "Ecumene" refers to land on which people dwell and make a livelihood; from ancient Greek, meaning the

habitable part of the earth. Pollard, *Marginal Europe,* especially applies here.

6 I was interested to discover W.A. Mackintosh referring to an "internal frontier," by which he apparently meant the dry margin: "Pioneer Belts Canadian Committee. 1929–31. Correspondence," Mackintosh to Isaiah Bowman, 23 Dec. 1931, in American Geographical Society archives, New York. Also Wonders, William C., "Marginal Settlement," *Scottish Geographical Magazine,* vol. 91 (1975), 12–24.

7 McKnight, Tom L., *Physical Geography: A Landscape Appreciation,* Englewood Cliffs: Prentice-Hall, 1984, 252–3, 256–8, 264–8.

8 Spence, C.C., "Land Utilization in Southwest Central Saskatchewan," *The Economic Annalist,* Dec. 1936, 84–8; Craig, G.H., and J. Proskie, "The Acquisition of Land in the Vulcan-Lomond Area of Alberta," *The Economic Annalist,* Oct. 1937, 68–74.

9 Richtik, James M., "Mapping the Quality of Land for Agriculture in Western Canada," *Great Plains Quarterly,* 5 (1985), 236–48. Shannon, E.N., "An Evaluation of the Physical Resources of the Meadow Lake Region," in R.G. Ironside, V.B. Proudfoot, E.N. Shannon, and C.J. Tracie, eds, *Frontier Settlement,* Edmonton: Department of Geography, University of Alberta, 1974, 136–50.

10 Tracie, Carl J., "Land of Plenty or Poor Man's Land. Environmental Perception and Appraisal Respecting Agricultural Settlement in the Peace River Country, Canada," in Brian W. Blouet and M.P. Lawson, eds, *Images of the Plains. The Role of Human Nature in Settlement,* Lincoln: University of Nebraska Press, 1975, 115–22; John Warkentin, ed., *The Western Interior of Canada. A Record of Geographical Discovery 1612–1917,* Toronto: McClelland & Stewart, 1964.

11 Canada, Department of Energy, Mines and Resources, *National Atlas,* "Growing Degree-Days" map; Ash, G.H.B., Shaykewich, C.F., and Raddatz, R.L., *Agricultural Climate of the Eastern Canadian Prairies,* Environment Canada, Manitoba Agriculture, and University of Manitoba.

12 Buchanan, Elizabeth, "In Search of Security: Kinship and the Farm Family on the North Shore of Lake Huron (Ontario), 1879–1939," PhD dissertation, McMaster University, 1989, chapters 3 and 4.
13 Gosselin, A., and G.-P. Boucher, *Problèmes de la Colonisation dans le nord du Nouveau-Brunswick*, Canada, Department of Agriculture, Economics Division, Publication no. 764 (Technical Bulletin no. 51), 1945, 6 (my translation).
14 Chapman, L.J., "The Climate of Northern Ontario," *Canadian Journal of Agricultural Science*, 33 (1953), 41.
15 Chapman, L.J., and M.K. Thomas, *The Climate of Northern Ontario*, Toronto: Canada, Department of Transport, Meteorological Branch, Climatological Studies no. 6, 1968, table 4.
16 Toews, Report to Premier Ferguson [1926].
17 Canada, *National Atlas*, "Growing Degree-Days"; Dunbar, Gary S., "Isotherms and Politics: Perception of the Northwest in the 1850's," in A.W. Rasporich and H.C. Klassen, eds, *Prairie Perspectives 2. Selected Papers of the Western Canadian Studies Conferences, 1970, 1971*, Toronto: Holt, Rinehart & Winston, 1973, 80–101.
18 Murchie and Grant, *Unused Lands*, 21–6.
19 Ash et al., *Agricultural Climate*, 4, figures 2, 5, and 8, analyses sixty years of records (from 1929) but does not necessarily give a more accurate picture than Murchie and Grant of the 1920s and 30s, because averages from the longer record smooth down extremes.
20 Executive Committee of Atlas of Alberta (J. Klawe, cartographic ed.), *Atlas of Alberta*, Edmonton: University of Alberta Press, 1969, 15.
21 Stone, Donald N.G., "Alberta's and British Columbia's Crown Lands Policies (1931–1973): Some Attitudinal and Behavioural Responses by Frontier Agriculturalists Towards Those Policies," PhD dissertation, University of Saskatchewan, 1980, 54–9 (quote from B.C. Lands Service, on 56).
22 Carder, A.C., *Climate of the Upper Peace River Region*, Canada, Department of Agriculture, Pub. 1224, Ottawa: Queen's Printer, 1965.

23 Carder, *Climate*, 9.
24 Ehlers, Eckart, *Das nördliche Peace River Country, Alberta, Kanada. Genese und Struktur eines Pionierraumes im borealen Waldland Nordamerikas*, Tübinger Geographische Studien, No. 18, Tübingen: Geographischen Instituts der Universität Tübingen, 1965, tables 2 and 3.
25 Ash et al., *Agricultural Climate*, 4, figures 3 and 6.
26 Murchie and Grant, *Unused Lands*, 21–2, maps 5, 6, 7, 8. For comparison, Carder, A.C., *Climatic Aberrations and the Farmer: Weather Extremes at Beaverlodge*, Beaverlodge: Canada Agriculture Research Station, 1967.
27 Fung, Ka-iu, ed., *Atlas of Saskatchewan*, Saskatoon: University of Saskatchewan, 1999, 96–7.
28 Stapleford, E.W., *Report on Rural Relief Due to Drought Conditions and Crop Failures in Western Canada 1930–1937*, Ottawa: Canada, Department of Agriculture, 1939, 26–7.
29 My analysis is based on Ash et al., *Agricultural Climate*, 7–9, figure 29.
30 Ash et al., *Agricultural Climate*, 8, figure 31.
31 Ash et al., *Agricultural Climate*, 8–9, figures 29, 30, 33.
32 Executive Committee (J. Klawe, cartographic ed.), *Atlas of Alberta*, 16.
33 Carder, *Climate*, 7.
34 Carder, *Climate*, figure 1; Ehlers, *Das nördliche*, table 2.
35 Elliott, G.C., "A Study of Wheat Yields in South-Central Saskatchewan," *Economic Annalist*, 8 (June 1938), 35–40; in Australia, a long-term yield of less than 6 bushels per acre was thought to be clearly below economic viability even with rail transport near: see Perkins, A.J., "Our Wheat-Growing Areas – Profitable and Unprofitable," *Journal of the Department of Agriculture [South Australia]*, 39 (1935–36), 1199–222.
36 Lehr, J.C., "The Rural Settlement Behaviour of Ukrainian Pioneers in Western Canada, 1891–1914," in B.M. Barr, ed., *Western Canadian Research in Geography: The Lethbridge Papers*, B.C. Geographical Series, no. 21, Vancouver: Tantalus Research 1975, 51–66.
37 Quoted in Fernow, *Conditions*, 5.

38 Hills, G.A., letter to J. Coke, Senior Economist, Economics Division, Canada Department of Agriculture, Ottawa, 19 Apr. 1944: in National Archives [hereafter NA], RG 17–3655, Canada Dep't of Agriculture, file N-6–22 "Colonization – Northern Ont. & Que."
39 "A Comparison of Earnings on the Black and Grey-Wooded Soils in Alberta," *Economic Annalist*, 14 (May 1944), 47.
40 Fernow, *Conditions*, 4–6.
41 Troughton, Michael J., "The Failure of Agricultural Settlement in Northern Ontario," *Nordia*, 17 (1983), 149–50; Gentilcore, R. Louis, and C. Grant Head, *Ontario's History in Maps*, Toronto: University of Toronto Press, 1984, 110 (maps 4.36, 4.37, 4.38); McDermott, George L., "Frontiers of Settlement in the Great Clay Belt, Ontario and Québec," *Annals, Association of American Geographers*, vol. 51 (1961), 262 (fig. 2).
42 Troughton, "Failure," 144–5; idem, "Persistent Problems of Rural Development in the 'Marginal Areas' of Canada," *Nordia*, vol. 12 (1978), 98.
43 McDermott, "Frontiers," 202–3.
44 *Soil Survey of Milk River Sheet* (1941), quoted in Flower, David, "Survival and Adaptation: An Analysis of Dryland Farming in the 1940s and 1950s in Southeast Alberta," PhD thesis, University of Alberta, 1997, 61.
45 Fung, *Atlas*, 220.
46 Hope, E.C., " Remarks on the Conditions in the Municipality of Shamrock, No. 134," 24 July 1935, typescript in University of Saskatchewan Archives, College of Agriculture Dean's Correspondence, II.B.55 (1935).
47 See evidence of the massive shift of government-assisted settlers in MacPherson, Murdo, "Drought and Depression on the Prairies," plate 43, in Donald Kerr and Deryck W. Holdsworth, eds, *Historical Atlas of Canada, Vol. 3: Addressing the Twentieth Century*, Toronto: University of Toronto Press, 1990.
48 Bailey, Mrs A.W., "The Year We Moved," *Saskatchewan History*, 20 (1967), 19.

49 Canada, Dominion Bureau of Statistics, *Agriculture, Climate and Population of the Prairie Provinces of Canada. A Statistical Atlas Showing Past Development and Present Conditions*, Ottawa: King's Printer, 1931, 78.

50 See gripping descriptions in Frederick Philip Grove's *Over Prairie Trails* [1922], Toronto: McClelland & Stewart, 1957, and Sinclair Ross's "The Painted Door," in his *The Lamp at Noon and Other Stories*, Toronto: McClelland & Stewart, 1968; Carder, *Climate*, 8.

51 Carder, *Climatic Aberrations*, 7.

## CHAPTER FOUR

1 McNeill, William H., *The Global Condition. Conquerors, Catastrophes, and Community*, Princeton: Princeton University Press, 1992, xiv.

2 Re slogans and labels, see Carl L. Becker, *The Heavenly City of the Eighteenth-Century Philosophers*, New Haven: Yale University Press, 1932, especially chapter 2.

3 Oleskow, Dr. Joseph, *On Emigration*, Publications of Michael Kachkowsky Society, Dec. 1895 – no. 241, Lviv, 1895, 4, 12, 14, in Public Archives of Alberta, Acc. 73.560 (translated from Ukrainian; said to be the only known copy on the American continent). Lehr, J.C., "Rural Settlement Behaviour of Ukrainian Pioneers in Western Canada, 1891–1914," in Brenton M. Barr, ed., *Western Canadian Research in Geography: The Lethbridge Papers*, B.C. Geographical Series, no. 21, Vancouver: Tantalus Research, 1975, 51–66.

4 Calculated from Canada, *Census of 1921*, vol. 2, table 32.

5 Calculated from Canada, *Census of 1931*, vol. 2, table 21.

6 MacPherson, Murdo, "Drought and Depression," plate 43 in Kerr and Holdsworth, eds, *Historical Atlas of Canada*, vol. 3.

7 Thompson, John H., "Bringing in the Sheaves: The Harvest Excursionists, 1890–1929," *Canadian Historical Review*, 59 (1978), 467–89.

8 Gosselin and Boucher, *Problèmes*; De Vries and MacNab-De Vries, *"They Farmed, Among Other Things."*

9 Turner, *Frontier and Section*, chapters 3, 4, 5; Tuttle, "Forerunners"; Savage and Thompson, "Comparative Study"; Powell, *Griffith Taylor*; Wood, "Limits reaffirmed."
10 Androchowiez, Julius, testimony to Saskatchewan, Royal Commission on Immigration and Settlement, in Dr William Swanson (Chair) papers, A4 – Box 2, Saskatchewan Archives Board, Saskatoon, typescript vol. 35, Prince Albert, 25 Apr. 1930, 26–7.
11 Collins testimony, Saskatchewan, Royal Commission on Immigration, 1930, vol. 6, Nipawin, 19 Feb. 1930, 21, 22, 29–30 (italics added); re choice of land, Lehr, "Rural Settlement Behaviour."
12 Grove, Frederick P., *Fruits of the Earth*, New Canadian Library, no. 49, Toronto: McClelland & Stewart, 1965 [1933], 17, 22.
13 Biays, Pierre, *Les Marges de L'Oekumene Dans L'Est du Canada*, Québec: L'Université Laval, 1964, 259 (my translation); N. Séguin, ed., *Agriculture et Colonisation.*
14 This discussion on Québec colonization in the 1920s and 30s is largely based on Courville, "Campagnes Inachevées," 8–14. The terms and expectations of the Québec system of colonization are discussed in the interviews with pioneers in Normand Lafleur, *La vie quotidienne des premiers colons en Abitibi-Témiscamingue*, Ottawa: Éditions Lémeac, 1976.
15 Canada, *Statutes*, "An Act respecting Relief Measures," 22–23 Geo. V, 1932, c. 36, and "An Act respecting Unemployment and Farm Relief," 22–23 Geo. V, 1932, c. 13. These and related acts are consolidated under subject headings in *The Canada Statute … Citator 1951.*
16 Courville, "Campagnes Inachevées," 9–10.
17 Courville, "Campagnes Inachevées," 10–11.
18 Courville, "Campagnes Inachevées," 11.
19 Gosselin, A., and G.P. Boucher, *Settlement Problems in Northwestern Quebec and Northeastern Ontario*, Ottawa: Canada, Department of Agriculture, Publication no. 758, 13, 17, and related, preliminary reports by Gosselin and Boucher in *The Economic Annalist*, Dec. 1938 (84–8), and in 1939, Feb. (7–11), Apr. (24–9), June (35–7), Aug. (56–61); see also Murdo

MacPherson, Serge Courville, and Daniel MacInnes, "Colonization and Co-operation," plate 44, in Kerr and Holdsworth, eds, *Historical Atlas of Canada*, vol. 3.

20 Lower, *Settlement and the Forest Frontier*, 57 (based on Annual Report of the Commissioner of Crown Lands for Canada, 1865). Re Québec-Ontario comparisons, see Frederick M. Helleiner and Guy Perrault, "Comparisons between Northeastern Ontario and Northwestern Québec" and Robert S. Dilley, "Farming on the Margin: Agriculture in Northern Ontario" in Margaret E. Johnston, ed., *Geographic Perspectives on the Provincial Norths*, Thunder Bay: Lakehead University, Centre for Northern Studies, Northern and Regional Studies Series, vol. 3, 1994, 163–79, 180–98.

21 Gosselin and Boucher, *Problèmes.*

22 Wynn, *Timber Colony*, 76–83; Roach, Thomas R., "The Pulpwood Trade and the Settlers of New Ontario, 1919–1938," *Journal of Canadian Studies*, vol. 22, no. 3 (1987), 78–88; Lower, *Settlement and the Forest Frontier*, 31, 48–57.

23 Biays, *Les Marges*, 276, (my translation); but for qualification of this, see Ian M. Drummond, ed., *Progress Without Planning: The Economic History of Ontario from Confederation to the Second World War*, Toronto: University of Toronto Press, 1987, 45–9. For a northwest Canada-Finland comparison, see Eckart Ehlers, "Recent Trends and Problems of Agricultural Colonization in Boreal Forest Lands," in R.G. Ironside et al., eds, *Frontier Settlement*, 63–6.

24 Sinclair, Peter W., "Agricultural Colonization in Ontario and Québec: Some Evidence from the Great Clay Belt, 1900–45," in D.H. Akenson, ed., *Canadian Papers in Rural History*, vol. v, Gananoque, ON: Langdale Press, 1986, 104–20.

25 In Griffin, Gerald, "Settler's Wife 'Never Dreamed Such Misery Existed in World' As Experienced in Northland," *Toronto Daily Star*, 6 June 1933.

26 "Political Motives Ascribed to Windsor Mayor's Exposure of Alleged Sufferings," *Cochrane Northland Post*, 8 June 1933.

27 Gosselin and Boucher, *Settlement Problems*, 17–21.

28 Owram, Doug, *Promise of Eden: The Canadian Expansionist Movement and the Idea of the West, 1856–1900*, Toronto:

University of Toronto Press, 1980, 223; see also Morton, W.L., "The Bias of Prairie Politics," in Swainson, Donald, ed., *Historical Essays on the Prairie Provinces*, Carleton Library, no. 53, Toronto: McClelland & Stewart, 1970, 289–300.

29 Sifton, Clifford, in *The Winnipeg Free Press*, 26 Feb. 1923, quoted in Lehr, "Rural Settlement Behaviour," 53.

30 Fedorowich, *Unfit for heroes*, chapter 3.

31 Fedorowich, *Unfit for heroes*, 65.

32 Murchie and Grant, *Unused Lands*, 60–1.

33 Murchie and Grant, *Unused Lands*, 61.

34 England, *Colonization*, chapter 5.

35 Weekly report from P.H. Ferguson, agricultural representative, to Saskatchewan's Deputy Minister of Agriculture, 15 Oct. 1921, in Saskatchewan Archives Board, Regina, R-259.II.8.

36 Report from J.F. Booth, Shaunavon, Sask., to Saskatchewan Deputy Minister of Agriculture, 5 July 1919, in Saskatchewan Archives Board, Regina, R-261.22.9.

37 England, *Colonization*, 107.

38 Stapleford, *Rural Relief*, 83–90, 75.

39 *The Sun*, Swift Current, Sask., 23 Oct. 1934.

40 From a committee reporting to the federal government, quoted in Bowen, Dawn, "'Forward to a Farm': Land Settlement as Unemployment Relief in the 1930s," *Prairie Forum*, 20 (1995), 215.

41 Stapleford, *Rural Relief*, 32, 39–43 (tables 6, 4, 6-C).

42 Stapleford, *Rural Relief*, 42 (tables 6-A, 6-B, 6-C).

43 Except as noted, the figures in this paragraph are from Stapleford, *Rural Relief*, especially 40–2 and 62–3; the quote, 39.

44 Saskatchewan, Attorney General, *A Submission by the Government of Saskatchewan to the Royal Commission on Dominion-Provincial Relations (Canada, 1937)*, 171, 181–7; Stapleford, *Rural Relief*, (for Alberta), 77 (table 11).

45 *Report of the Survey Board for Southern Alberta*, Edmonton: King's Printer, 1922, 7.

46 MacLean, Una D., "The Special Areas of Alberta: An Historical Survey," typescript, a Glenbow Foundation

Project (in Glenbow Museum and Archives, Calgary), 1959, 53–5.

47 This translation and fuller details in McCallum, Charlotte, "Québec's Reactions to the 1920 Manitoba Mennonite Search for Land," *Journal of Mennonite Studies*, vol. 20 (2002), 43–57 (quote 51–2). For context see Epp, Frank H., *Mennonites in Canada, 1920–1940. A People's Struggle for Survival*, Toronto: Macmillan, 1982, 109–13.

48 England, *Colonization*, 137, 141.

49 Friesen, *Canadian Prairies*, 267–70.

50 Bowen, Dawn, "*Die Auswanderung*: religion, culture, and migration among Old Colony Mennonites," *Canadian Geographer*, vol. 45 (2001), 461–73.

51 Friesen, *Canadian Prairies*, 267, 269.

52 Tracie, *"Toil and Peaceful Life"*; Friesen, *Canadian Prairies*, 269–70.

53 Clark, "Huterian Communities," 357–60.

54 Peters, Victor, "The Hutterians: History and Communal Organization of a Rural Group," in Donald Swainson, ed., *Historical Essays on the Prairie Provinces*, Carleton Library, no. 53, Toronto: McClelland & Stewart, 1970, 41, 131 (quote).

55 Waddington, Miriam, "Memoirs of a Jewish Farmer," *NeWest ReView*, (Sept. 1980) 5–7.

56 Katz and Lehr, "Jewish and Mormon Agricultural Settlement," 129–33.

57 Painchaud, Robert, "French-Canadian Historiography and Franco-Catholic Settlement in Western Canada, 1870–1915," *Canadian Historical Review*, 59 (1978), 447–66.

58 Normandeau, The Rev. J.-A., *L'Alberta Centrale*, Montreal: np, 1914; Michaud, Georges, *L'Avenir Agricole des Canadiens français en Saskatchewan*, np: np, 1928 (my translation of quote from second page of text).

59 Biays, *Les Marges*, 259; Morissonneau, Christian, *La Terre promise: Le mythe du Nord québécois*, Montréal: Hurtubise, 1978, espec. 26–30; Courville, "Campagnes Inachevées," 10–11.

60 Québec, Ministry of Colonization, Mines and Fisheries, *Les Régions de Colonisation de la Province de Québec. La Mattavinie*, Québec, 1920, frontispiece.

61 Séguin, *Agriculture et Colonisation*, 27–37, quote 29 (my translation).
62 Semaines Sociales du Canada, "Programme," *Congrès de la Colonisation ... 1944 à Montréal. Compte Rendu*, Montréal: École Sociale Populaire.
63 Saskatchewan, Royal Commission on Immigration and Settlement, Cabri, 6 Mar. 1930, vol. 16, 23–6, Saskatchewan Archives Board, Regina, R-249.
64 Glazebrook, G.P. de T., *A History of Transportation in Canada*, vol. 2, National Economy 1867–1936, Carleton Library, no. 12, Toronto: McClelland & Stewart, 1964, chapter 10, 66–7 (map).
65 England, *Colonization*, 74; Morissonneau, *La Terre Promise*, 179–82.
66 See these maps in Mackintosh, W.A., *Prairie Settlement. The Geographical Setting*, Canadian Frontiers of Settlement, vol. 1, Toronto: Macmillan Co., 1934, 48–52.
67 Hedges, James B., *Building the Canadian West. The Land and Colonization Policies of the Canadian Pacific Railway*, New York: Russell & Russell, [1939] 1971, chapter 9.
68 Thompson, "Bringing in the Sheaves," especially table 1.
69 Hedges, *Building the Canadian West*, 315–16.
70 Hedges, *Building the Canadian West*, 332–6.
71 Hedges, *Building the Canadian West*, chapter 12.
72 Osborne and Wurtele, "The Other Railway," 235–44.
73 Glazebrook, *History of Transportation*, chapter 11.
74 Osborne and Wurtele, "The Other Railway," 236–8, 241.
75 Androchowiez testimony, Saskatchewan, Royal Commission on Immigration and Settlement, Prince Albert, 25 April 1930, vol. 35, 28.
76 Brown, Ralph H., *Historical Geography of the United States*, New York: Harcourt, Brace & World, 1948; Harris, R. Cole, and Warkentin, John, *Canada Before Confederation. A Study in Historical Geography*, Ottawa: Carleton University Press, 1991; Robert D. Mitchell and Paul A. Groves, eds, *North America: The Historical Geography of a Changing Continent*, Totowa, NJ: Rowman & Littlefield, 1987.
77 Laut, "The Last Trek," 102–3.

78 Zaslow, Morris, *The Northward Expansion of Canada 1914–1967*, Toronto: McClelland & Stewart, 1988, 31.

## CHAPTER FIVE

1 From quotations in Langford, "First Generation," 95, 49, 84.
2 Interview with Mr and Mrs William H. Howes (here Mr Howes), Porcupine Plain settlement, Saskatchewan Archives Board, Saskatoon, C 19, 19a.
3 Jones, *Empire of Dust*; Rees, Ronald, *New and Naked Land. Making the Prairies Home*, Saskatoon: Western Producer Prairie Books, 1988.
4 Hart, J. Fraser, "The Spread of the Frontier and the Growth of Population," in H.J. Walker and W.G. Haag, eds, *Man and Culture*, vol. 5 of *Geoscience & Man* (1974), 73–81; Barrows, Harlan H., *Geography of the Middle Illinois Valley*, Illinois State Geological Survey, Bull. no. 15, Urbana: University of Illinois, 1910.
5 Brunger, Alan G., "Geographical Propinquity among Pre-Famine Catholic Irish Settlers in Upper Canada," *Journal of Historical Geography*, 8 (1982), 265–82; Malin, James, "The Turnover of Farm Population in Kansas," *Kansas Historical Quarterly*, vol. 4 (1935), 339–72.
6 Danysk, *Hired Hands*, chapter 5; idem, "A Bachelor's Paradise," in Cavanaugh and Mouat, eds, *Making Western Canada*, 154–85.
7 Danysk, *Hired Hands*, 76.
8 MacEwan, Grant, *Fifty Mighty Men*, Saskatoon: Modern Press, 1958, 169–74.
9 Quoted in Langford, "First Generation," 106.
10 *Le Livre du Colon ou "Comment s'installer sur une terre pour presque rien,"* Connaissance des Pays Québécois/Patrimoine, 1979 [orig.1902], reissued by *La Terre de chez nous* (newspaper), Longueuil, 1996.
11 *Le Livre du Colon*, 5–7.
12 Laliberté, Joseph, with Laplante, Robert, *Agronome-Colon en Abitibi*, Québec: Institut Québécois de Recherche sur la Culture, 1983, 25, 26, 27, 31, 32. The population of

Roquemaure was 486 in 1931 and approximately double that at the opening of 1941.

13 Quoted in "Back-to-Land Families Are Moving South," *Northern News*, Kirkland Lake, 27 April 1933, A1.

14 Grayson, L.M., and Bliss, Michael, eds, *The Wretched of Canada. Letters to R.B. Bennett 1930–1935*, Toronto: University of Toronto Press, 1971, 72, 56.

15 Gosselin and Boucher, *Settlement Problems*, 52–3.

16 Quoted in "The Other Side Of Settlers' Problems," *Cochrane Northland Post*, 8 June 1933, 23.

17 Jackson, Dr Mary Percy, *Suitable for the Wilds: Letters from Northern Alberta, 1929–1931*, edited with introduction by Janice Dickin McGinnis, Toronto: University of Toronto Press, 1995.

18 Jackson, *Suitable for the Wilds*, 177–8, 190–1.

19 Alberta, Department of Attorney General, Alberta Provincial Police Annual Reports, Provincial Archives of Alberta, Acc. No. 72.370, Peace River subdistrict, 1930, 19.

20 South African scrip was a reward offered to veterans of the Boer War, redeemable for Crown land in Canada (normally 320 acres). It was commonly sold by veterans for a small sum.

21 It was the mid-1940s before more than half of the farmers in the Prairie Provinces owned tractors: Ankli, Robert E., Helsberg, H. Dan, Thompson, John H., "The Adoption of the Gasoline Tractor in Western Canada," in Donald Akenson, ed., *Canadian Papers in Rural History*, vol. 2 (1980), 10.

22 From taped interview of Mr and Mrs Frank Kinderwater, formerly of the La Glace, Alberta, area, conducted in Grande Prairie, Alberta, c. 1965 by Dr Carl Tracie, who has kindly permitted use of this material.

23 Kerpan, Gabriel, typescript of interview in Parksville, B.C., by daughter Mary, Oct. 1991. Re depressed wheat prices, see MacPherson, Ian, and Thompson, J.H., "An orderly Reconstruction: Prairie Agriculture in World War Two," in Akenson, ed., *Canadian Papers in Rural History*, vol. 4, Gananoque: Langdale Press, 1984, 11–13.

24 Langford, "First Generation," 15. See this strongly championed in Buss, Helen M., "Settling the Score with Myths of Settlement: Two Women Who Roughed It and Wrote It," (and other essays) in Elspeth Cameron and Janice Dickin, eds, *Great Dames*, Toronto: University of Toronto Press, 1997, 167–83. For many related references, see the excellent bibliography in Cavanaugh and Mouat, *Making Western Canada*, 278.
25 Margaret Thompson (1919), quoted in Langford, "First Generation," 33–4; G. Chase (1922), *idem*, 106.
26 Quoted in Langford, "First Generation," 82, 86.
27 Excerpted from "Hail on the Homestead," by Irene Louise Harrison, in *From Out of the Wilderness. A History of Baptiste Lake … West Athabasca, and Winding Trail School Districts*, Public Archives of Alberta, #971.233, at 3201.
28 Quoted in Langford, "First Generation," 88.
29 Mary Tennis (1926) quoted in Langford, "First Generation," 167.
30 Androchowiez testimony, Saskatchewan, Royal Commission on Immigration and Settlement, Prince Albert, 25 April 1930, vol. 35, 23, 24, 26, 28.
31 Collins testimony, Saskatchewan, Royal Commission on Immigration and Settlement, Nipawin, 19 Feb. 1930, 22–31.
32 Murchie and Grant, *Unused Lands*, 64; Dawson, *Group Settlement*, 377; Vanderhill, Burke G., "Settlement in the Forest Lands of Manitoba, Saskatchewan, and Alberta: A Geographic Analysis," doctoral dissertation, University of Michigan, 1956, 203; Rice, John G., "The role of culture and community in frontier prairie farming," *Journal of Historical Geography*, vol. 3 (1977), 155–75.
33 Enns, John H., "The Story of the Mennonite Settlement of Reesor, Ontario," typescript, 1973 (in the Ron Morel Memorial Museum, Kapuskasing, ON); *Regina Leader-Post*, 9 May 1934.
34 Courville, "Campagnes Inachevées," 17–18.
35 Harris, Herbert R., *Book of Memories and a History of the Porcupine Soldier Settlement and Adjacent Areas Situated in*

*North-Eastern Saskatchewan Canada 1919–1967*, Shand Agricultural Society, 1967, 5.

36 Rouse, Wayne R., "Climatic Change and Water in the Arctic Waterheds of Manitoba and Ontario," in Margaret E. Johnston, ed., *Provincial Norths*, 339.

37 Ontario, *Statutes*, 7–8 Geo. V, c. 13 (1917), "An Act Providing for the Agricultural Settlement of Soldiers and Sailors Serving Overseas"; Ontario, Commission of Inquiry, "Report Commission of Enquiry Kapuskasing Colony, 1920," *Sessional Papers*, vol. 52, part VIII, 1920, 3–4.

38 Vincent, Odette, ed., *Histoire de l'Abitibi-Témiscamingue*, Institut québécois de recherche sur la culture, Ste-Foy: Les Presses de l'Université Laval, 1995, 229.

39 Adams, Thomas, *Rural Planning and Development. A Study of Rural Conditions and Problems in Canada*, Ottawa: Commission of Conservation, Canada, 1917, 207.

40 Canada, 7–8 Geo. V, c. 21 (1917), "An Act to assist Returned Soldiers in settling upon the Land and to increase Agricultural production" (revised and elaborated in 1919 by c. 71).

41 Kirkconnell, Watson, "Kapuskasing – An Historical Sketch," *Queen's Quarterly*, vol. 28 (Jan. 1921), 264–78; Zaslow, *Northward Expansion of Canada 1914–1967*, Toronto: McClelland & Stewart, 1988, 43–5.

42 Excerpts from Commission of Inquiry, "Kapuskasing Colony ..."; some parts have been repositioned to improve flow and integration. The report has thirteen pages of text. For political context, see Johnston, Charles M., *E.C. Drury: Agrarian Idealist*, Toronto: University of Toronto Press, 1986, 166–71.

43 Fernow, *Conditions*, 11, appendix 1; Troughton, "Failure," also notes the lack of a plan appropriate for the physical base and the social needs.

44 Kirkconnell, W., "Kapuskasing," 277; Murchie, R.W., *Land Settlement as a Relief Measure*, University of Minnesota Press, 1933, 15; Lower, *Settlement and the Forest Frontier*, 144. For a thorough review of population change and mention of some ethnicity-based settlements, see Wightman, Robert and Nancy, "Changing Patterns of Rural Peopling in Northeastern Ontario, 1901–1941," *Ont. History*, vol. 92 (2000), 161–81.

45 Quoted in Harris, *Book of Memories*, 10.
46 Harris, *Book of Memories*, 7–11. Zaslow, Morris, *The Northward Expansion*, 41, describes the Kapuskasing soldier settlement as "clearly ... a failure," but the Porcupine Plain scheme, a little overstated, "completely successful."
47 Clarence Hassard in Harris, *Book of Memories*, 12–15.
48 Fedorowich, *Unfit for Heroes*, 198, 104–5.
49 Murchie and Grant, *Unused Lands*, 61; Murchie, *Land Settlement*, 15.
50 Adams, *Rural Planning*, 209, 210.
51 Powell, "The Debt of Honour," 72.
52 Troper, Harold Martin, "The Creek-Negroes of Oklahoma and Canadian Immigration, 1909–11," *Canadian Historical Review*, vol. 53 (1972), 272–88. The racial/ethnic terminology has modulated from Negroes or "colored folks," as the Amber Valley people called themselves in the interviews, to the current general acceptance of "blacks."
53 Statutes of Canada, 9–10 Edw. VII (1909–10), c. 27, "The Immigration Act," s.38 (c); Shepard, R. Bruce, *Deemed Unsuitable. Blacks from Oklahoma Move to the Canadian Prairies in Search of Equality*, Toronto: Umbrella Press, 1997, 100; Grow, Stewart, "The Blacks of Amber Valley – Negro Pioneering in Northern Alberta," *Canadian Ethnic Studies*, vol. 6 (1974), 17–38. Apparently the order-in-council of Aug. 1911 was not explicitly put into effect.
54 Grow, "The Blacks of Amber Valley," from which most of the detail and interview material in this case study is taken.
55 Excerpt from Mr J.D. Edwards interview in Grow, "The Blacks of Amber Valley," 29, 32.
56 Excerpt from Edwards interview in Grow, "The Blacks of Amber Valley," 30.
57 Excerpts from interviews with Edwards and Thomas Mapp, in Grow, "The Blacks of Amber Valley," 31–2.
58 Excerpt from interview with Mrs J.D. Edwards, in Grow, "The Blacks of Amber Valley," 33.
59 Ringuet, [Philippe Panneton], *Thirty Acres*, New Canadian Library, Toronto: McClelland & Stewart, 1989, 20.

60 Katz and Lehr, "Jewish and Mormon Agricultural Settlement," 140.
61 Waddington, "Memoirs," 6.
62 Katz and Lehr, "Jewish and Mormon Agricultural Settlement," 138.
63 Waddington, "Memoirs," 6, 7.
64 England, *Colonization*, 275; Katz and Lehr, "Jewish and Mormon Agricultural Settlement," 132.
65 Toews, Report to Premier Ferguson.
66 Enns, "The Story of … Reesor."
67 Dyck, Mrs Peter (neé Erna Toews), quoted in Enns, "The Story of … Reesor," 8, 9, 10–11, 15.
68 Enns, "The Story of … Reesor," 103.
69 This account of La Crete is drawn from Bowen, *"Die Auswanderung"*; physical background from Executive Committee and Klawe, *Atlas of Alberta*.
70 For examples of the technique, see Katz and Lehr, "Jewish and Mormon Agricultural Settlement," figures 4, 5; Carlyle, William J., Lehr, John C., Mills, G.E., "Peopling the Prairies," plate 17 in Kerr and Holdsworth, eds, *Historical Atlas of Canada*, vol. 3; J. David Wood, "Scandinavian Settlers in Canada Revisited," *Geografiska Annaler*, Ser. B, vol. 49 (1967), 1–9. For domesticated space, see Courville, Serge, "Space, Territory and Culture in New France: A Geographical Perspective," in Graeme Wynn, ed., *People, Places, Patterns, Processes: Geographical Perspectives on the Canadian Past*, Toronto: Copp, Clark, Pitman, 1990, 165–76.
71 Johnson, Mr and Mrs G.R., *The Northfield Settlement 1913–1969*, Grande Prairie, AB: Printed by Menzies Printers, 1969.
72 Rackham, Thomas S., *Back To The Land. A Study of the Dominion-Provincial Rehabilitation Plan in Alberta 1932–48*, Ottawa: Canada Dept of Agriculture/Alberta Dept of Agriculture, 1953.
73 Rackham, *Back To The Land*, 6.
74 Rackham, *Back To The Land*, 1–12.
75 Rackham, *Back To The Land*, 6, 30–1.
76 Stutt, R.A., "Average Progress of Settlers in the Albertville-Garrick, Northern Pioneer Areas, Saskatchewan, 1941," *Economic Annalist*, Aug. 1943, 44–50.

77 Stutt, "Average Progress," 50.
78 Bowman, "The Pioneer Fringe," 49.
79 Murchie and Grant, *Unused Lands*, 191, (my italics).
80 Bowen, "Forward to a Farm"; Harris, *Book of Memories*; Commission of Inquiry, "Kapuskasing Colony."
81 Zaslow, *Northward Expansion*, 44–9.
82 Troughton, "Failure," 143–9.
83 Bélanger, Marcel, "Le Québec rural," in Fernand Grenier, ed., *Québec*, Toronto: University of Toronto Press, 1972, 35–41. Note that French Canadians continued to be successful in expanding farm settlement in Ontario's clay belts: Wightman, "Changing Patterns."
84 Allen, William, "Farm Management Problems in the Prairie Provinces of Western Canada," *Farm Economics*, Cornell University, no. 67 (Aug. 1930), 1361–5; "The Decline in Prices," *Economic Annalist*, vol. 11 (Jan. 1932), 1.
85 Grant, H.C., "Taxation in Pioneer Areas of Manitoba," appendix C in Mackintosh, *Economic Problems*, 296–7.

## CHAPTER SIX

1 Fernow, *Conditions*, 11.
2 "Pioneer Experiences" questionnaires, Saskatchewan Archives Board, Howes interview re Porcupine Plain, C 19, 19.
3 Stapleford, *Rural Relief*, Section VII; Canada, Department of Agriculture, *P.F.R.A. A Record of Achievement*, A Report ... for the Eight-Year Period ended March 31, 1943. Community Pastures are mapped in MacPherson, "Drought and Depression," in Kerr and Holdsworth, eds, *Historical Atlas of Canada*, vol. 3, plate 43.
4 Vanderhill, "Settlement in the Forest Lands," chapter 4; idem., "The Decline of Land Settlement in Manitoba and Saskatchewan," *Economic Geography*, vol. 38 (1962), 270–77; idem., "The ragged edge: A review of contemporary agricultural settlement along the Canadian northern frontier," *Geografisch Tijdschrift*, vol. 5 (1971), 123–33.
5 Mackintosh, W.A., et al., *Economic Problems of the Prairie Provinces*, Canadian Frontiers of Settlement, vol. IV, Toronto:

Macmillan, 1935, 234; for a pan-Canadian commentary, see Troughton, "Persistent Problems," 97–107.

6 Bowman, "The Scientific Study" [1926], and "Introduction" in I. Bowman, ed., *Limits of Land Settlement. A Preliminary Report of Present-day Possibilities*, New York: American Coordinating Committee for International Studies, 1937.

7 Bélanger, "Le Québec rural," 46; Laliberté, *Agronome-Colon*.

8 Martin, Chester, *"Dominion Lands" Policy*, by Lewis H. Thomas, ed., Carleton Library no. 69, Toronto: McClelland & Stewart, 1973, 240.

9 Bowen, *"Die Auswanderung."*

10 Stone, Kirk H., "Geographic Aspects of Planning for New Rural Settling in the Free World's Northern Lands," in Saul B. Cohen, ed., *Problems and Trends in American Geography*, New York: Basic Books, 1967, 221–38.

11 Vanderhill, "The ragged edge."

12 Eggleston, "Old Homestead," 127.

13 Voisey, Paul, "A Mix-up over Mixed Farming: The Curious History of the Agricultural Diversification Movement in a Single Crop Area of Southern Alberta," in David C. Jones and Ian MacPherson, eds, *Building Beyond the Homestead: Rural History on the Prairies*, Calgary: University of Calgary Press, 1985, 179–205.

14 Flower, "Survival and Adaptation," 235–44.

15 Eggleston, "Old Homestead," 127; MacLean, "Special Areas," re Tilley East Area Act (1927) followed by the Berry Creek Area Act and others a few years later.

16 Stutt, "Average Progress," 46–7; Acton, B.K., "A Comparison of Farms in the Grande Prairie District of Alberta 1930 and 1942," *Economic Annalist*, Aug. 1943, 53–6.

17 Ehlers, Eckart, *Das nördliche Peace River Country, Alberta, Kanada. Genese und Struktur eines Pionierraumes im borealen Waldland Nordamerikas*, Tübinger Geographische Studien, No. 18, Tübingen: Universität Tübingen, 1965, 103–5, 130–2.

18 Zaslow, *Northward Expansion*, 44–5.

19 Leuthold, Frank O., "Communication and Diffusion of Improved Farm Practices in Two Northern Saskatchewan Farm Communities," Saskatoon: Canadian Centre for

Community Studies, 1966, 14. Shannon's study of the Meadow Lake district in northwestern Saskatchewan adds details; for example, between 1941 and 1971 the number of farm operators decreased precipitously but total land in farms went up by 30 per cent, and improved land tripled; and the numbers of all farm animals went down except for the tripling of non-dairy cattle; Shannon, "An Evaluation of the Physical Resources," 136–40.

20 Vanderhill, Burke G., "Post-War Agricultural Settlement in Manitoba," *Economic Geography*, vol. 35 (1959), 261–5.

21 Vanderhill, "The Decline," 274–6.

22 Vanderhill, "The Decline," 276, and "The ragged edge," 127–30.

23 Voisey, *Vulcan*, 98–114.

24 Stutt, R. A., "Changes in the Extent and Effect of Mechanization on Saskatchewan Farms," *Economic Annalist*, Aug. 1944, 57–62.

25 Ankli, Helsberg, and Thompson, "The Adoption of the Gasoline Tractor," 9.

26 Flower, "Survival and Adaptation," 235–42.

27 MacPherson and Thompson, "An Orderly Reconstruction," 11–32.

28 Davies, I.G., "Agriculture in the Northern Forest – the Case of Northwestern Ontario," *The Lakehead University Review*, vol. 1 (1968), 129–53.

29 Bentley, C.F., "Soil Management and Fertility – Western Canada," in Canada, Department of Northern Affairs and National Development, *Resources for Tomorrow. Conference Background Papers*, vol. 1, July 1961, 69.

30 Friesen, *Canadian Prairies*, 389–92.

31 Troughton, "The Failure," 143–6; Helleiner and Perrault, "Comparisons," 173–7.

32 Dilley, "Farming on the Margin," 186–8; Shannon, "An Evaluation of the Physical Resources," 136–40.

# Bibliography

## PRIMARY SOURCES, INCLUDING NEWSPAPERS

Alberta, Department of Attorney General, Alberta Provincial Police Annual Reports, Provincial Archives of Alberta, Acc. No. 72.370, Peace River subdistrict, 1930.

Alberta, Executive Council, Orders in Council 641/34, in Provincial Archives of Alberta, Acc. No. 70.427 (1934).

Androchowiez, Julius, testimony to Saskatchewan, Royal Commission on Immigration and Settlement, in Dr William Swanson (Chair) papers, A4-Box 2, Saskatchewan Archives Board, Saskatoon, typescript vol. 35, Prince Albert, 25 Apr. 1930.

"Back-to-Land Families Are Moving South," *Northern News*, Kirkland Lake, 27 April 1933, A1.

Baker, O.E., "Memorandum for Dr. Bowman, Chairman Committee on Pioneer Belts," 31 July 1926, in University of Saskatchewan Archives, Saskatoon: College of Agriculture, Dean's Correspondence – Canadian Pioneer Belts Project, 1928–31.

Booth, J.F., Report to Saskatchewan Deputy Minister of Agriculture, 5 July 1919, in Saskatchewan Archives Board, Regina, R-261.22.9.

"Climat Abitibien, Le," *La Terre de Chez Nous*, 20 Nov. 1935.

Collins, A., testimony to Saskatchewan, Royal Commission on Immigration and Settlement, 1930, typescript, vol. 6, Nipawin, 19 Feb. 1930, Saskatchewan Archives Board, Saskatoon, Dr William Swanson (Chair) papers, A4-Box 1.

Cunningham, T.D., letter to Premier Aberhart, 18 Nov. 1936, in Premiers' Papers, file 2280A, Alberta Provincial Archives.

"Decline in Prices, The," *Economic Annalist* 11 (Jan. 1932), 1.

Enns, John H., "The Story of the Mennonite Settlement of Reesor, Ontario," typescript, 128 pp., 1973 (in the Ron Morel Memorial Museum, Kapuskasing, ON).

"Femme du Colon, La" (reprinted from *La Canadienne* magazine), *L'Abitibi*, 22 Apr. 1920.

Ferguson, P.H., Weekly report to Saskatchewan Deputy Minister of Agriculture, 15 Oct. 1921, in Saskatchewan Archives Board, Regina, R-259.II.8.

Griffin, Gerald, "Settler's Wife 'Never Dreamed Such Misery Existed in World' As Experienced in Northland," *Toronto Daily Star*, 6 June 1933.

Hills, G.A., letter to J. Coke, Senior Economist, Economics Division, Canada Department of Agriculture, Ottawa, 19 Apr. 1944: in National Archives [hereafter NA], RG 17-3655, Canada Department of Agriculture, file N-6-22, "Colonization – Northern Ont. & Que."

Hope, Prof. E.C., "Remarks on the Conditions in the Municipality of Shamrock, No. 134," 24 July 1935, typescript in University of Saskatchewan Archives, College of Agriculture Dean's Correspondence, II.B.55 (1935).

Howes, Mr and Mrs William H., Porcupine Plain settlement, transcript of tape-recorded interview, Saskatchewan Archives Board, Saskatoon, C 19.

Kerpan, Gabriel, typescript of interview in Parksville, B.C., by daughter, Oct. 1991.

Kinderwater, Mr and Mrs Frank, formerly of the La Glace, Alberta, area, from taped interview by Dr Carl Tracie (in Grande Prairie, Alberta, c. 1965).

Mackintosh, W.A., Letter to Isaiah Bowman, 23 Dec. 1931, "Pioneer Belts Canadian Committee. 1929–31. Correspondence," in American Geographical Society archives, New York.

MacLean, Una D., "The Special Areas of Alberta: An Historical Survey," typescript, A Glenbow Foundation Project (in Glenbow Museum and Archives, Calgary), 1959.

Minute book of Shaunavon District Home Makers Club, 16 Sept. 1937, Saskatchewan Archives Board, Regina, R-1151.

"Not Adapted as Settlers," editorial, *Sudbury Star*, 29 Apr. 1933.

"Other Side Of Settlers' Problems, The," *Cochrane Northland Post*, 8 June 1933, 23.

"Political Motives Ascribed to Windsor Mayor's Exposure of Alleged Sufferings," *Cochrane Northland Post*, 8 June 1933.

Saskatchewan, Deputy Minister of Agriculture (Auld), letter to A.E. MacNab, Secretary of Bruce Preparedness League, Walkerton, ON, 20 June 1918, in Saskatchewan Archives Board, Regina: R-261, Department of Agriculture, Deputy Minister files, 26.5 – Soldier Settlements.

Saskatchewan, Royal Commission on Immigration and Settlement, testimonies, Cabri, 6 Mar. 1930, vol. 16, Saskatchewan Archives Board, Regina, R-249; also, Nipawin, 19 Feb. 1930, vol. 6, and Prince Albert, 25 Apr. 1930, vol. 35, in Dr William Swanson (Chair) papers, A4 – boxes 1,2, Saskatchewan Archives Board, Saskatoon.

Stenhouse, Mrs Peter, Letter to Saskatchewan Minister of Agriculture, 3 Apr. 1936, in Saskatchewan Department of Agriculture, R-261, 2.229, Saskatchewan Archives Board, Regina.

Toews, Jacob, Report, in Ontario Archives, Premier Howard Ferguson – Correspondence received, RG 03-06-0-1289 (reel MS 1721).

## SECONDARY SOURCES, INCLUDING THESES

Acton, B.K., "A Comparison of Farms in the Grande Prairie District of Alberta 1930 and 1942," *Economic Annalist*, Aug. 1943, 53–6.

Adams, Thomas, *Rural Planning and Development. A Study of Rural Conditions and Problems in Canada*, Ottawa: Commission of Conservation, Canada, 1917.

Alberta, *Report of the Survey Board for Southern Alberta*, Edmonton: King's Printer, 1922.

Albright, W.D., "An Economic Pioneer Land Settlement Policy," CSTA *Review,* No. 35 (Dec. 1942).

Allen, William, "Farm Management Problems in the Prairie Provinces of Western Canada," *Farm Economics,* Cornell University, No. 67 (Aug. 1930), 1361–5.

Anderson, Alan B. "Ethnic Identity in Saskatchewan Bloc Settlements: A Sociological Appraisal," in Howard Palmer, ed., *The Settlement of the West.* Calgary: University of Calgary, 1977, 187–225.

Ankli, Robert E., H. Dan Helsberg, and John H. Thompson, "The Adoption of the Gasoline Tractor in Western Canada," in Donald Akenson, ed., *Canadian Papers in Rural History* 2 (1980), 9–39.

Ash, G.H.B., C.F. Shaykewich, and R.L. Raddatz, *Agricultural Climate of the Eastern Canadian Prairies.* Environment Canada and Manitoba Agriculture, 1992.

Badé, William Frederic, *The Life and Letters of John Muir.* Boston: Houghton Mifflin, 1924.

Bailey, Mrs A.W., "The Year We Moved," *Saskatchewan History* 20 (1967), 19–31.

Baker, O.E., "Government Research in Aid of Settlers and Farmers in the Northern Great Plains States of the United States," in W.L.G. Joerg, ed., *Pioneer Settlement. Cooperative Studies by Twenty-Six Authors.* New York: American Geographical Society, Special Publication No. 14, 1932, 61–79.

Barron, Hal S., *Those Who Stayed Behind: Rural Society in Nineteenth-Century New England.* Cambridge: Cambridge University Press, 1984.

Barrows, Harlan H., *Geography of the Middle Illinois Valley.* Illinois State Geological Survey, Bull. No. 15. Urbana: University of Illinois, 1910.

Beard, Charles A., "The Idea of Progress," in Charles A. Beard, ed., *A Century of Progress.* New York: Harper, 1932.

Becker, Carl L., *The Heavenly City of the Eighteenth-Century Philosophers.* New Haven: Yale University Press, 1932.

Bélanger, Marcel, "Le Québec rural," in Fernand Grenier, ed., *Québec.* Toronto: University of Toronto Press, 1972, 35–41.

Bentley, C.F., "Soil Management and Fertility – Western Canada," in Canada, Department of Northern Affairs and National

Development, *Resources for Tomorrow. Conference Background Papers*, vol. 1, July 1961.

Biays, Pierre, *Les Marges de L'Oekumene Dans L'Est du Canada*. Québec: L'Université Laval, 1964.

Bouchard, Gérard, "Family Reproduction in New Rural Areas: Outline of a North American Model," *Canadian Historical Review* 75 (1994), 475–510.

Boulding, Kenneth E., "The Economics of the Coming Spaceship Earth," in *Beyond Economics. Essays on Society, Religion, and Ethics*. Ann Arbor: University of Michigan Press, 1968.

Bowen, Dawn, "*Die Auswanderung*: religion, culture, and migration among Old Colony Mennonites," *Canadian Geographer* 45 (2001), 461–73.

– "'Forward to a Farm': Land Settlement as Unemployment Relief in the 1930s," *Prairie Forum* 20 (1995), 207–29.

Bowen, William A., "Mapping an American Frontier: Oregon in 1850," Map Supplement No. 18, *Annals of the Association of American Geographers* 65 (Mar. 1975).

Bowman, Isaiah. "Introduction," in American Co-ordinating Committee for International Studies, *Limits of Land Settlement. A Preliminary Report on Present-day Possibilities*, U.S. Memorandum No. 2, 1937.

– ed., *Limits of Land Settlement. A Preliminary Report of Present-day Possibilities*. New York: American Coordinating Committee for International Studies, 1937.

– "The Pioneer Fringe," *Foreign Affairs* 6 (1927), 49–66.

– "The Scientific Study of Settlement," *Geographical Review* 16 (1926), 647–53.

Brown, Ralph H. *Historical Geography of the United States*. New York: Harcourt, Brace & World, 1948.

Brunger, Alan G., "Geographical Propinquity among Pre-Famine Catholic Irish Settlers in Upper Canada," *Journal of Historical Geography* 8 (1982), 265–82.

Buchanan, Elizabeth, "In Search of Security: Kinship and the Farm Family on the North Shore of Lake Huron (Ontario), 1879–1939," PhD dissertation, McMaster University, 1989.

Buss, Helen M., "Settling the Score with Myths of Settlement: Two Women Who Roughed It and Wrote It," in Elspeth Cameron and Janice Dickin, eds, *Great Dames.* Toronto: University of Toronto Press, 1997, 167–83.

Canada, "An Act to assist Returned Soldiers in settling upon the Land and to increase Agricultural production," 7–8 Geo. V, c. 21 (1917) (revised and elaborated in 1919 by c. 71).

Canada, "An Act respecting Unemployment and Farm Relief," 22–23 Geo. V, 1932, c. 13.

Canada, "An Act respecting Relief Measures," 22–23 Geo. V, 1932, c. 36.

Canada, "The Immigration Act," s.38 (c), 9–10 Edw. VII (1909–10), c. 27.

Canada, *Census of 1921.*

Canada, *Census of 1931.*

Canada, *Census of 1941.*

Canada, *Census of The Prairie Provinces, 1926.*

Canada, *Census of The Prairie Provinces, 1936.*

Canada, Dominion Bureau of Statistics, *Agriculture, Climate and Population of the Prairie Provinces of Canada. A Statistical Atlas Showing Past Development and Present Conditions.* Ottawa: King's Printer, 1931.

Canada, House of Commons, *Official Report of the Debates,* 5 Edw. VII (1905), vol. I. Ottawa: King's Printer, 1905.

Canada, Department of Agriculture, P.F.R.A. *A Record of Achievement,* A Report ... for the Eight-Year Period ended March 31, 1943.

Canada, Department of Energy, Mines and Resources, *The National Atlas of Canada,* 5th edition. Ottawa: Geographical Services Directorate, 1981.

Canada, Royal Commission on the Natural Resources of Saskatchewan, *Report.* Ottawa: King's Printer, 1935.

Carder, A.C., *Climatic Aberrations and the Farmer: Weather Extremes at Beaverlodge.* Beaverlodge: Canada Agriculture Research Station, 1967.

– *Climate of the Upper Peace River Region.* Canada, Department of Agriculture, Pub. 1224. Ottawa: Queen's Printer, 1965.

Carlyle, William J., John C. Lehr, and G.E. Mills, "Peopling the Prairies," plate 17 in D. Kerr and D. Holdsworth, eds, *Historical Atlas of Canada,* vol. 3. Toronto: University of Toronto Press, 1990.

Carson, Rachel, *Silent Spring* [1962]. Harmondsworth: Penguin, 1965.

Chapman, L.J., "The Climate of Northern Ontario," *Canadian Journal of Agricultural Science* 33 (1953), 47–73.

– and M.K. Thomas, *The Climate of Northern Ontario*. Toronto: Canada, Department of Transport, Meteorological Branch, Climatological Studies No. 6, 1968.

Clark, Bertha W., "The Huterian Communities," *Jour. of Political Economy* 32 (1924), 357–74, 468–86.

Cloke, Paul, and Jo Little, eds, *Contested Countryside Cultures: Otherness, Marginalisation and Rurality*. London: Routledge, 1997.

"Comparison of Earnings on the Black and Grey-Wooded Soils in Alberta, A," *Economic Annalist* 14 (May 1944).

Courville, Serge, "Des Campagnes Inachevées: L'Exemple du Nord Québécois," in Brian S. Osborne, ed., *Canada's Countryside* (forthcoming).

– "Space, Territory and Culture in New France: A Geographical Perspective," in Graeme Wynn, ed., *People, Places, Patterns, Processes: Geographical Perspectives on the Canadian Past*. Toronto: Copp, Clark, Pitman, 1990, 165–76.

Craig, G.H., and J. Proskie, "The Acquisition of Land in the Vulcan-Lomond Area of Alberta," *The Economic Annalist*, Oct. 1937, 68–74.

Curti, Merle. *The Making of An American Community: A Case Study of Democracy in a Frontier County*. Stanford: Stanford University Press, 1959.

Danysk, Cecilia. "'A Bachelor's Paradise': Homesteaders, Hired Hands, and the Construction of Masculinity, 1880–1930," in Catherine Cavanaugh and Jeremy Mouat, eds, *Making Western Canada. Essays on European Colonization and Settlement*. Toronto: Garamond Press, 1996, 154–85.

– *Hired Hands: Labour and the Development of Prairie Agriculture, 1880–1930*. Toronto: McClelland & Stewart, 1995.

Davies, I.G., "Agriculture in the Northern Forest – the Case of Northwestern Ontario," *The Lakehead University Review* 1 (1968), 129–53.

Dawson, C.A., *Group Settlement. Ethnic Communities in Western Canada*, vol. 7 of the Canadian Frontiers of Settlement series. Toronto: Macmillan, 1936.

– "The Social Structure of a Pioneer Area as Illustrated by the Peace River District," in W.L.G. Joerg, ed., *Pioneer Settlement. Cooperative Studies by Twenty-Six Authors*. New York: American Geographical Society, Special Publication No. 14, 1932, 37–49.

– and Eva Younge, *Pioneering in the Prairie Provinces: The Social Side of the Settlement Process*, Canadian Frontiers of Settlement, vol. 8. Toronto: Macmillan, 1940.

De Blij, H.J., Peter O. Muller, and Richard S. Williams, Jr, *Physical Geography: The Global Environment*, 3rd edition. New York/Oxford: Oxford University Press, 2004.

De Vries, Pieter J. and Georgina MacNab-De Vries. *"They Farmed, Among Other Things." Three Cape Breton Case Studies*. Sydney, NS: University College of Cape Breton Press, 1983.

Dilley, Robert S., "Farming on the Margin: Agriculture in Northern Ontario," in Margaret E. Johnston, ed., *Geographic Perspectives on the Provincial Norths*. [Thunder Bay]: Lakehead University, Centre for Northern Studies, Northern and Regional Studies Series, vol. 3, 1994, 180–98.

Drummond, Ian M., ed., *Progress Without Planning: The Economic History of Ontario from Confederation to the Second World War*. Toronto: University of Toronto Press, 1987.

Dunbar, Gary S. "Isotherms and Politics: Perception of the Northwest in the 1850's," in A. W. Rasporich and H. C. Klassen, eds, *Prairie Perspectives 2. Selected Papers of the Western Canadian Studies Conferences, 1970, 1971*. Toronto: Holt, Rinehart & Winston, 1973, 80–101.

Easterlin, Richard A. "Population Change and Farm Settlement in the Northern United States," *JEcHist* 36 (Mar. 1976), 45–75.

Eggleston, Wilfrid, "The Old Homestead: Romance and Reality," in Howard Palmer, ed., *The Settlement of the West*. Calgary: University of Calgary, Comprint Publishing Co., 1977.

Ehlers, Eckart, *Das nördliche Peace River Country, Alberta, Kanada. Genese und Struktur eines Pionierraumes im borealen Waldland Nordamerikas*, Tübingen Geographische Studien, No. 18. Tübingen: Geographischen Instituts der Universität Tübingen, 1965.

– "Recent Trends and Problems of Agricultural Colonization of Boreal Forest Lands," in R.G. Ironside, V.B. Proudfoot, E.N. Shannon, and C.J. Tracie, eds, *Frontier Settlement*. Edmonton: University of Alberta Press, 1974, 60–78.

Elliott, G.C., "A Study of Wheat Yields in South-Central Saskatchewan," *Economic Annalist* 8 (June 1938), 35–40.

England, Robert, *The Colonization of Western Canada. A Study of Contemporary Land Settlement (1896–1934)*. London: King & Son, 1936.

– *The Central European Immigrant in Canada*. Toronto: Macmillan, 1929.

Epp, Frank H. *Mennonites in Canada, 1920–1940. A People's Struggle for Survival*. Toronto: Macmillan, 1982.

Executive Committee of Atlas of Alberta (J. Klawe, cartographic ed.), *Atlas of Alberta*. Edmonton: University of Alberta Press, 1969.

Faucher, Albert, "Explication socio-économique des migrations dans l'histoire du Québec," in Normand Séguin, ed., *Agriculture et Colonisation au Québec: Aspects historiques*. Montréal: Boréal Express, 1980, 141–58.

Fedorowich, Kent, *Unfit for heroes. Reconstruction and Soldier Settlement in the Empire between the Wars*. Manchester: Manchester University Press, 1995.

Fernow, B.E., *Conditions in the Clay Belt of New Ontario*. Ottawa: Commission of Conservation, 1912.

Fleure, H.J., "Human Regions," *Scottish Geographical Magazine* 35 (1919), 94–105.

Flower, David, "Survival and Adaptation: An Analysis of Dryland Farming in the 1940s and 1950s in Southeast Alberta," unpublished PhD thesis, University of Alberta, 1997.

Friesen, Gerald, *The Canadian Prairies. A History*. Toronto: University of Toronto Press, 1984.

Fung, Ka-iu, ed., *Atlas of Saskatchewan*. Saskatoon: University of Saskatchewan, 1999.

Gagan, David, *Hopeful Travellers: Families, Land, and Social Change in Mid-Victorian Peel County, Canada West*. Toronto: University of Toronto Press, 1980.

– and H. Mays, "Historical Demography and Canadian Social History: Families and Land in Peel County, Ontario," *Canadian Historical Review* 54 (1973), 27–47.

Gentilcore, R. Louis, and C. Grant Head, *Ontario's History in Maps*. Toronto: University of Toronto Press, 1984.

Georgeson, C.C., "The Possibilities of Agricultural Settlement in Alaska," in W.L.G. Joerg, ed., *Pioneer Settlement. Cooperative*

*Studies by Twenty-Six Authors*. New York: American Geographical Society, Special Publication No. 14, 1932, 50–60.

Girard, Michel F., *L'écologisme rétrouvé. Essor et déclin de la Commission de la conservation du Canada*. Ottawa: Les Presses de l'Université d'Ottawa, 1994.

Glacken, Clarence J. *Traces on the Rhodian Shore. Nature and Culture in Western Thought from Ancient Times to the End of the Eighteenth Century*. Berkeley and Los Angeles: University of California Press, 1967.

Glazebrook, G.P. de T., *A History of Transportation in Canada*, vol. II, National Economy 1867–1936, Carleton Library No. 12. Toronto: McClelland & Stewart, 1964.

Gosselin, A., and G.-P. Boucher, *Problèmes de la Colonisation dans le nord du Nouveau-Brunswick*, Canada, Department of Agriculture, Economics Division, Publication No. 764 (Technical Bulletin No. 51), 1945.

– *Settlement Problems in Northwestern Québec and Northeastern Ontario*, Ottawa: Canada, Department of Agriculture, Publication No. 758.

Grant, H.C., "Taxation in Pioneer Areas of Manitoba," Appendix C in W.A. Mackintosh, et al., *Economic Problems of the Prairie Provinces*, Canadian Frontiers of Settlement, vol. 4. Toronto: Macmillan, 1935, 296–7.

Grayson, L.M., and Michael Bliss, eds, *The Wretched of Canada. Letters to R. B. Bennett 1930–1935*. Toronto: University of Toronto Press, 1971.

Grove, Frederick P., *Settlers of the Marsh* [1925], New Canadian Library 50. Toronto: McClelland & Stewart, 1966.

– *Fruits of the Earth* [1933], New Canadian Library 49. Toronto: McClelland & Stewart, 1965.

– *Over Prairie Trails* [1922]. Toronto: McClelland & Stewart, 1957.

Grow, Stewart, "The Blacks of Amber Valley – Negro Pioneering in Northern Alberta," *Canadian Ethnic Studies* 6 (1974), 17–38.

Guelph Conference on the Conservation of the Natural Resources of Ontario. *Conservation and Post-War Rehabilitation*, [n.p.], 1942.

Harris, Herbert R., *Book of Memories and a History of the Porcupine Soldier Settlement and Adjacent Areas Situated in North-Eastern Saskatchewan Canada 1919–1967*. Shand Agricultural Society, 1967.

Harris, R. Cole, and John Warkentin, *Canada Before Confederation. A Study in Historical Geography.* Ottawa: Carleton University Press, 1991.

Harrison, Irene Louise, "Hail on the Homestead," in *From Out of the Wilderness. A History of Baptiste Lake … West Athabasca, and Winding Trail School Districts*, Public Archives of Alberta, #971.233.

Hart, J. Fraser, "The Spread of the Frontier and the Growth of Population," in H.J. Walker and W.G. Haag, eds, *Man and Culture…*, vol. 5, *Geoscience & Man* (1974), 73–81.

Hedges, James B., *Building the Canadian West. The Land and Colonization Policies of the Canadian Pacific Railway.* New York: Russell & Russell [1939], 1971.

Helleiner, Frederick M., and Guy Perrault, "Comparisons Between Northeastern Ontario and Northwestern Québec," in Margaret E. Johnston, ed., *Geographic Perspectives on the Provincial Norths*, [Thunder Bay]: Lakehead University, Centre for Northern Studies, Northern and Regional Studies Series, vol. 3, 1994, 163–79.

Inge, William, "The Idea of Progress," The Romanes Lecture. Oxford: Clarendon Press, 1920.

Jackson, Dr Mary Percy, *Suitable for the Wilds: Letters from Northern Alberta, 1929–1931*, edited with introduction by Janice Dickin McGinnis. Toronto: University of Toronto Press, 1995.

Joerg, W.L.G., ed., *Pioneer Settlement. Cooperative Studies by Twenty-Six Authors*. New York: American Geographical Society, Special Publication No. 14, 1932.

Johnson, Mr and Mrs G.R., *The Northfield Settlement 1913–1969*. [Grande Prairie, AB: Menzies Printers, 1969].

Johnston, Charles M., *E. C. Drury: Agrarian Idealist*. Toronto: University of Toronto Press, 1986.

Jones, David C., *Empire of Dust: Settling and Abandoning the Prairie Dry Belt*. Edmonton: The University of Alberta Press, 1987.

Jussila, Heikki, Walter Leimgruber, and Roser Majoral, eds, *Perceptions of Marginality. Theoretical Issues and Regional Perceptions of Marginality in Geographical Space*. Aldershot: Ashgate Publishing Ltd, 1998.

Katz, Yossi, and John Lehr, "Jewish and Mormon Agricultural Settlement in Western Canada: A Comparative Analysis," *Can. Geographer* 35, no. 2 (1991), 128–42.

Kirkconnell, Watson, "Kapuskasing – An Historical Sketch," *Queen's Quarterly* 28 (Jan. 1921), 264–78.

Lafleur, Normand, *La vie quotidienne des premiers colons en Abitibi-Témiscamingue*. Ottawa: Editions Leméac, 1976.

Lake, Marilyn, *The Limits of Hope: Soldier Settlement in Victoria, 1915–38*. Melbourne: Oxford University Press, 1987.

Laliberté, Joseph, with Robert Laplante, *Agronome-Colon en Abitibi*. Québec: Institut Québécois de Recherche sur la Culture, 1983.

Langford, Nanci L., "First Generation and Lasting Impressions: The Gendered Identities of Prairie Homestead Women," unpublished PhD thesis, University of Alberta, 1994.

Laut, Agnes C., "The Last Trek to the Last Frontier. The American Settler in the Canadian Northwest," *Century Illustrated Monthly Magazine* 78 (1909), 99–111.

Lehr, J.C. "Rural Settlement Behaviour of Ukrainian Pioneers in Western Canada, 1891–1914," in Brenton M. Barr, ed., *Western Canadian Research in Geography: The Lethbridge Papers*, B.C. Geographical Series, No. 21. Vancouver: Tantalus Research, 1975, 51–66.

Leuthold, Frank O., "Communication and Diffusion of Improved Farm Practices in Two Northern Saskatchewan Farm Communities," Saskatoon: Canadian Centre for Community Studies, 1966.

*Livre du Colon, Le, ou "Comment s'installer sur une terre pour presque rien,"* Connaissance des Pays Québécois/Patrimoine, 1979 [orig. 1902], reissued by *La Terre de chez nous* (newspaper), Longueuil, 1996.

Lower, A.R.M., *Settlement and the Forest Frontier in Eastern Canada*, first part of Canadian Frontiers of Settlement, vol. 9. Toronto: Macmillan, 1936.

MacEwan, Grant, *Fifty Mighty Men*. Saskatoon: Modern Press, 1958, 169–74.

Mackintosh, W.A., *Prairie Settlement. The Geographical Setting*, Canadian Frontiers of Settlement, vol. I. Toronto: Macmillan Co., 1934.

– *et al.*, *Economic Problems of the Prairie Provinces*, Canadian Frontiers of Settlement, vol. 4. Toronto: Macmillan, 1935.

MacPherson, Ian, and John H. Thompson, "An Orderly Reconstruction: Prairie Agriculture in World War Two," in Donald Akenson, ed., *Canadian Papers in Rural History* 4 (1984), 11–32.

MacPherson, Murdo, "Drought and Depression on the Prairies," plate 43, in Donald Kerr and Deryck W. Holdsworth, eds, *Historical Atlas of Canada*, vol. 3: *Addressing the Twentieth Century*. Toronto: University of Toronto Press, 1990.

– Serge Courville, and Daniel MacInnes, "Colonization and Cooperation," plate 44, in Kerr and Holdsworth, eds, *Historical Atlas of Canada*, vol. 3: *Addressing the Twentieth Century*. Toronto: University of Toronto Press, 1990.

Malin, James, "The Turnover of Farm Population in Kansas," *Kansas Historical Quarterly* 4 (1935), 339–72.

Marsh, George Perkins, *Man and Nature; or, Physical Geography as Modified by Human Action*, [1864], David Lowenthal, ed. Cambridge, MA: Belknap Press, 1965.

Martin, Chester, *"Dominion Lands" Policy*, Lewis H. Thomas, ed., Carleton Library 69. Toronto: McClelland & Stewart, 1973.

Matthews, Geoffrey J., and Robert Morrow, Jr, *Canada and the World*, second edition. Scarborough: Prentice-Hall, 1995.

McCallum, Charlotte, "Québec's Reactions to the 1920 Manitoba Mennonite Search for Land," *Journal of Mennonite Studies* 20 (2002), 43–57.

McDermott, George L. "Frontiers of Settlement in the Great Clay Belt, Ontario and Québec," *Annals, Association of American Geographers* 51 (1961).

McInnis, R.M., "Childbearing and Land Availability: Some Evidence from Individual Household Data," in Ronald D. Lee, ed., *Population Patterns in the Past*. New York: Academic Press, 1977, 201–27.

McKnight, Tom L., *Physical Geography: A Landscape Appreciation*. Englewood Cliffs: Prentice-Hall, 1984.

McNeill, William H., *The Global Condition. Conquerors, Catastrophes, and Community*. Princeton: Princeton University Press, 1992.

Michaud, Georges, *L'Avenir Agricole des Canadiens français en Saskatchewan*. np: np, 1928.

Miller, David Harry, and Jerome O. Steffen, eds, *The Frontier. Comparative Studies*. Norman: University of Oklahoma Press, 1977.

Mitchell, Robert D. and Paul A. Groves, eds., *North America: The Historical Geography of a Changing Continent*. Totowa, NJ: Rowman & Littlefield, 1987.

Morissonneau, Christian. *La Terre promise: Le mythe du Nord québécois*. Montréal: Hurtubise, 1978.

Morton, W.L., "The Bias of Prairie Politics," in Donald Swainson, ed., *Historical Essays on the Prairie Provinces*, Carleton Library 53. Toronto: McClelland & Stewart, 1970.

Murchie, R.W., *Land Settlement as a Relief Measure*. University of Minnesota Press, 1933.

– and H.C. Grant, *Unused Lands of Manitoba. Report of Survey 1926*. Winnipeg: Manitoba, Department of Agriculture and Immigration, 1927.

Nelles, H.V. *The Politics of Development. Forests, Mines & Hydro-electric Power in Ontario, 1849–1941*. Toronto: Macmillan of Canada, 1974.

Normandeau, The Rev. J.-A., *L'Alberta Centrale*. Montréal: np 1914.

Norris, Darrell, "Household and Transiency in a Loyalist Township: The People of Adolphustown, 1784–1822," *Histoire Sociale/ Social History* 13, no. 26 (1980), 399–415.

Oleskow, Dr Joseph, *On Emigration*, Publications of the Michael Kachkowsky Society, Dec. 1895 – no. 241, Lviv, 1895, in Public Archives of Alberta, Acc. 73.560.

Ontario, "An Act Providing for the Agricultural Settlement of Soldiers and Sailors Serving Overseas," 7–8 Geo. V, c. 13 (1917).

Ontario, Commission of Inquiry, "Report Commission of Enquiry Kapuskasing Colony, 1920," *Sessional Papers*, vol. 52, part 8, 1920.

Ontario, Department of Agriculture, *Report on the Reforestation of Waste Lands in Southern Ontario, 1908*, by E.J. Zavitz. Toronto: King's Printer, 1909.

Osborne, Brian S., and Susan Wurtele, "The Other Railway: Canadian National's Department of Colonization and Agriculture," *Prairie Forum* 20 (1995), 231–53.

Ost, Lillian, ed., *Seven Persons. One Hundred Sixty Acres and a Dream*. Medicine Hat: Seven Persons Historical Society, 1982.

Owram, Doug, *Promise of Eden: The Canadian Expansionist Movement and the Idea of the West, 1856–1900*. Toronto: University of Toronto Press, 1980.

Painchaud, Robert, "French-Canadian Historiography and Franco-Catholic Settlement in Western Canada, 1870–1915," *Canadian Historical Review*, 59 (1978), 447–66.

Peale, Norman Vincent, *The Power of Positive Thinking for Young People*. Englewood Cliffs: Prentice-Hall, 1952.

Perkins, A.J., "Our Wheat-Growing Areas – Profitable and Unprofitable," *Journal of the Department of Agriculture [South Australia]* 39 (1935–36), 1199–222.

Peters, Victor, "The Hutterians: History and Communal Organization of a Rural Group," in Donald Swainson, ed., *Historical Essays on the Prairie Provinces*, Carleton Library 53. Toronto: McClelland & Stewart, 1970.

Pollard, Sidney, *Marginal Europe. The Contribution of Marginal Lands since the Middle Ages*. Oxford: Clarendon Press, 1997.

Powell, J.M., *Griffith Taylor and "Australia Unlimited,"* J.M. Macrossan Memorial Lecture, 13 May 1992. St Lucia: University of Queensland Press, 1993.

– *An Historical Geography of Modern Australia: The Restive Fringe*. Cambridge: Cambridge University Press, 1988.

– "The Debt of Honour: Soldier Settlement in the Dominions, 1915–1940," *Journal of Australian Studies* 8 (1981), 64–87.

Québec, Ministry of Colonization, Mines and Fisheries, *Les Régions de Colonisation de la Province de Québec. La Mattavinie*, Québec, 1920.

Rackham, Thomas S., *Back To The Land. A Study of the Dominion-Provincial Rehabilitation Plan in Alberta 1932–48*. Ottawa: Canada, Department of Agriculture/Alberta, Department of Agriculture, 1953.

Rae, Thomas I., *The Administration of the Scottish Frontier: 1513–1603*. Edinburgh: Edinburgh University Press, 1966.

Rees, Ronald, *New and Naked Land. Making the Prairies Home*. Saskatoon: Western Producer Prairie Books, 1988.

*Regina Leader-Post*, 9 May 1934.

Reitel, François, "Le Rôle de l'Armée dans la Conservation des Forêts en France," *Bulletin de l'Association de Géographes Français* 502–3 (1984), 143–54.

Rice, John G., "The Role of Culture and Community in Frontier Prairie Farming," *Jour. of Historical Geography* 3 (1977), 155–75.

Richtik, James M., "Mapping the Quality of Land for Agriculture in Western Canada," *Great Plains Quarterly* 5 (1985), 236–48.

Ringuet [Philippe Panneton], *Thirty Acres*, New Canadian Library. Toronto: McClelland & Stewart, 1989.

Roach, Thomas R. "The Pulpwood Trade and the Settlers of New Ontario, 1919–1938," *Journal of Canadian Studies* 22, no. 3 (1987), 78–88.

Roche, Michael, "Soldier Settlement in New Zealand after World War I: two case studies," *New Zealand Geographer* 58 (2002), 23–32.

– "Fit Land for Heroes: A Reappraisal of Discharged Soldier Settlement in New Zealand after World War I," paper presented to the International Conference of Historical Geography, Université Laval, Ste Foy, Québec, August 2001.

Ross, Sinclair, *The Lamp at Noon and Other Stories*. Toronto: McClelland & Stewart, 1968.

Rouse, Wayne R., "Climate, Climatic Change and Water in the Arctic Watersheds of Manitoba and Ontario," in Margaret E. Johnston, ed., *Geographic Perspectives on the Provincial Norths*. [Thunder Bay]: Lakehead University Centre for Northern Studies, Northern and Regional Studies Series, vol. 3, 1994, 337–57.

Saskatchewan, Attorney General, *A Submission by the Government of Saskatchewan to the Royal Commission on Dominion-Provincial Relations (Canada, 1937)*. [Regina: King's Printer, 1937].

Savage, William W., Jr, and Stephen I. Thompson, "The Comparative Study of the Frontier: An Introduction," in William W. Savage and Stephen I. Thompson, eds, *The Frontier. Comparative Studies*. Norman: University of Oklahoma Press, 1979.

Séguin, Normand, "L'histoire de l'agriculture et de la colonisation au Québec depuis 1850," in Normand Séguin, ed., *Agriculture et Colonisation au Québec: Aspects historiques*. Montréal: Boréal Express, 1980, 9–38.

Semaines Sociales du Canada, "Programme," *Congrès de la Colonisation ... 1944 à Montréal. Compte Rendu*. Montréal: École Sociale Populaire.

Semple, Neil, *The Lord's Dominion: The History of Methodism in Canada.* Montreal, Kingston: McGill-Queen's University Press, 1996.

Shannon, E.N., "An Evaluation of the Physical Resources of the Meadow Lake Region [Saskatchewan]," in R.G. Ironside, V.B. Proudfoot, E.N. Shannon, and C.J. Tracie, eds, *Frontier Settlement.* Edmonton: University of Alberta Press, 1974, 130–50.

Shepard, R. Bruce, *Deemed Unsuitable. Blacks from Oklahoma Move to the Canadian Prairies in Search of Equality ...* Toronto: Umbrella Press, 1997.

Shields, Rob, *Places on the Margin: Alternative Geographies of Modernity.* London: Routledge, 1991.

Sinclair, Peter W., "Agricultural Colonization in Ontario and Québec: Some Evidence from the Great Clay Belt, 1900–45," in D.H. Akenson, ed., *Canadian Papers in Rural History* 5. Gananoque, ON: Langdale Press, 1986, 104–20.

Spence, C.C., "Land Utilization in Southwest Central Saskatchewan," *The Economic Annalist*, Dec. 1936, 84–8.

Stapleford, E.W., *Report on Rural Relief Due to Drought Conditions and Crop Failures in Western Canada 1930–1937.* Ottawa: Canada, Department of Agriculture, 1939.

Stone, Donald N.G., "Alberta's and British Columbia's Crown Lands Policies (1931–1973): Some Attitudinal and Behavioural Responses by Frontier Agriculturalists Towards Those Policies." PhD dissertation, University of Saskatchewan, 1980.

Stone, Kirk H., "Geographic Aspects of Planning for New Rural Settling in the Free World's Northern Lands," in Saul B. Cohen, ed., *Problems and Trends in American Geography.* New York: Basic Books, 1967, 221–38.

Stutt, R.A., "Changes in the Extent and Effect of Mechanization on Saskatchewan Farms," *Economic Annalist*, Aug. 1944, 57–62.

– "Average Progress of Settlers in the Albertville-Garrick, Northern Pioneer Areas, Saskatchewan, 1941," *Economic Annalist*, Aug. 1943, 44–50.

Taylor, Griffith, "The Frontiers of Settlement in Australia," *Geographical Review* 16 (1926), 1–6.

Thompson, J.H., "An Orderly Reconstruction: Prairie Agriculture in World War Two," in D.H. Akenson, ed., *Canadian Papers in Rural History* 4. Gananoque: Langdale Press, 1984, 11–15.

– "Bringing in the Sheaves: The Harvest Excursionists, 1890–1929," *Canadian Historical Review* 59 (1978), 467–89.

Tracie, Carl J., *"Toil and Peaceful Life." Doukhobor Village Settlement in Saskatchewan, 1899–1918*. Regina: Canadian Plains Research Centre, University of Regina, 1996.

– "Land of Plenty or Poor Man's Land. Environmental Perception and Appraisal Respecting Agricultural Settlement in the Peace River Country, Canada," in Brian W. Blouet and M.P. Lawson, eds, *Images of the Plains. The Role of Human Nature in Settlement*. Lincoln: University of Nebraska Press, 1975, 115–22.

Troper, Harold Martin, "The Creek-Negroes of Oklahoma and Canadian Immigration, 1909–11," *Canadian Historical Review* 53 (1972), 272–88.

Troughton, Michael J., "The Failure of Agricultural Settlement in Northern Ontario," *Nordia* 17 (1983), 141–51.

– "Persistent Problems of Rural Development in the 'Marginal Areas' of Canada," *Nordia* 12 (1978), 97–107.

Turner, F.J., "The Significance of the Frontier in American History," in *Frontier and Section. Selected Essays of Frederick Jackson Turner*. Englewood Cliffs: Prentice-Hall, 1961, 37–62.

Tuttle, William M., Jr, "Forerunners of Frederick Jackson Turner: Nineteenth Century British Conservatives and the Frontier Thesis," *Agricultural History* 41 (1967), 219–27.

Tyman, John Langton, *By Section, Township and Range. Studies in Prairie Settlement*. Brandon: Assiniboine Historical Society, 1972.

Vanderhill, Burke G., "The Ragged Edge: A Review of Contemporary Agricultural Settlement along the Canadian Northern Frontier," *Geografisch Tijdschrift* 5 (1971), 123–33.

– "The Decline of Land Settlement in Manitoba and Saskatchewan," *Economic Geography* 38 (1962), 270–7.

– "Post-War Agricultural Settlement in Manitoba," *Economic Geography* 35 (1959), 261–5.

– "Settlement in the Forest Lands of Manitoba, Saskatchewan, and Alberta: A Geographic Analysis." Doctoral dissertation, University of Michigan, 1956.

Vincent, Odette, ed., *Histoire de l'Abitibi-Témiscamingue*, Institut québécois de recherche sur la culture. Ste-Foy: Les Presses de l'Université Laval, 1995.

Voisey, Paul, *Vulcan: The Making of a Prairie Community.* Toronto: University of Toronto Press, 1988.

– "A Mix-up over Mixed Farming: The Curious History of the Agricultural Diversification Movement in a Single Crop Area of Southern Alberta," in David C. Jones and Ian MacPherson, eds, *Building Beyond the Homestead: Rural History on the Prairies.* Calgary: University of Calgary Press, 1985, 179–205.

Waddington, Miriam, "Memoirs of a Jewish Farmer," *NeWest ReView* (Sept. 1980), 5–7.

Warkentin, John, ed., *The Western Interior of Canada. A Record of Geographical Discovery 1612–1917.* Toronto: McClelland & Stewart, 1964.

Wightman, W. Robert, and Nancy M. Wightman, "Changing Patterns of Rural Peopling in Northeastern Ontario, 1901–1941," *Ont. History* 92 (2000), 161–81.

– *The Land Between: Northwestern Ontario Resource Development, 1800 to the 1990s.* Toronto: University of Toronto Press, 1997.

Williams, Michael, *The Making of the South Australian Landscape: A Study in the Historical Geography of Australia.* London, New York: Academic Press, 1974.

Wonders, William C. "Marginal Settlement," *Scottish Geographical Magazine* 91 (1975), 12–24.

Wood, David, "Limits Reaffirmed: New Wheat Frontiers in Australia, 1916–39," *Journal of Historical Geography* 23 (1997), 459–77.

– "The Population of Ontario: A Study of the Foundation of a Social Geography," in Guy M. Robinson, ed, *A Social Geography of Canada.* Toronto: Dundurn Press, 1991, 92–137.

– "Population Change on an Agricultural Frontier: Upper Canada, 1796 to 1841," in Roger Hall, W. Westfall, L.S. MacDowell, eds, *Patterns of the Past. Interpreting Ontario's History.* Toronto: Dundurn Press, 1988, 55–77.

– "Scandinavian Settlers in Canada Revisited," *Geografiska Annaler*, Ser. B, vol. 49 (1967), 1–9.

Wynn, Graeme, *Timber Colony: A Historical Geography of Early Nineteenth Century New Brunswick.* Toronto: University of Toronto Press, 1981.

Zaslow, Morris, *The Northward Expansion of Canada 1914–1967.* Toronto: McClelland & Stewart, 1988.

# Index